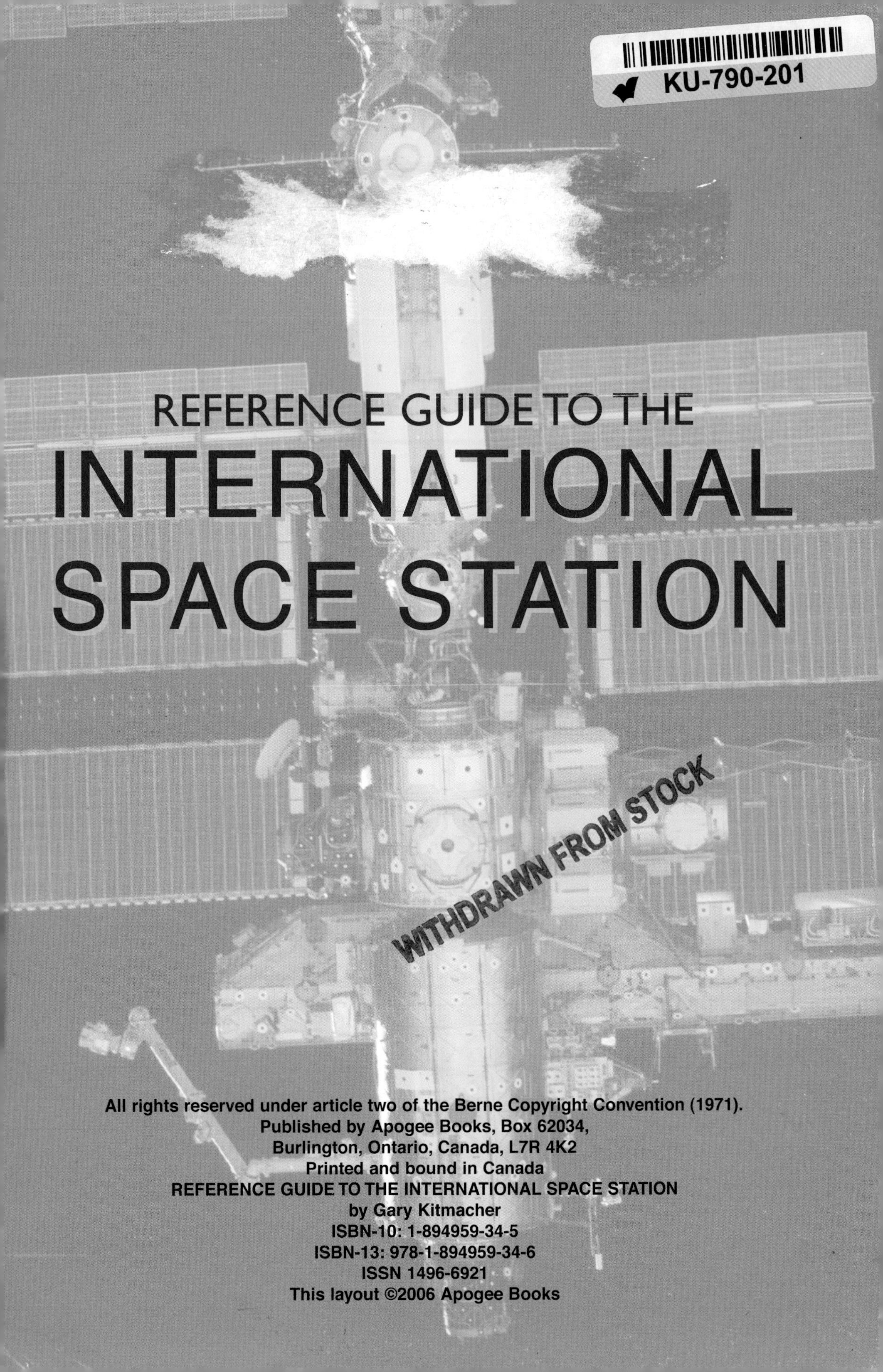

REFERENCE GUIDE TO THE INTERNATIONAL SPACE STATION

Published by Apogee Books, Box 62034,
Burlington, Ontario, Canada, L7R 4K2
Printed and bound in Canada
REFERENCE GUIDE TO THE INTERNATIONAL SPACE STATION
by Gary Kitmacher
ISBN-10: 1-894959-34-5
ISBN-13: 978-1-894959-34-6
ISSN 1496-6921

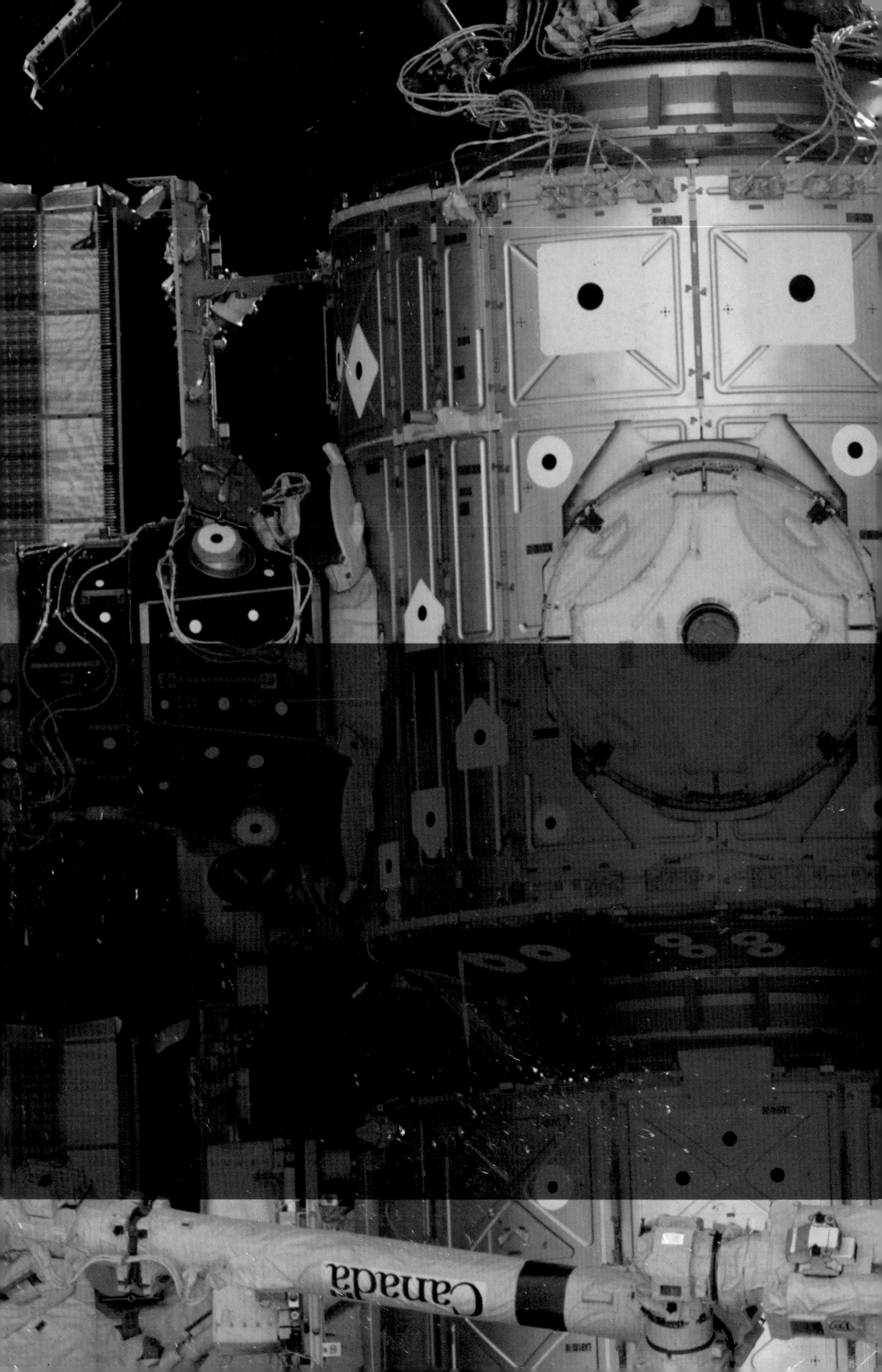
Canadä

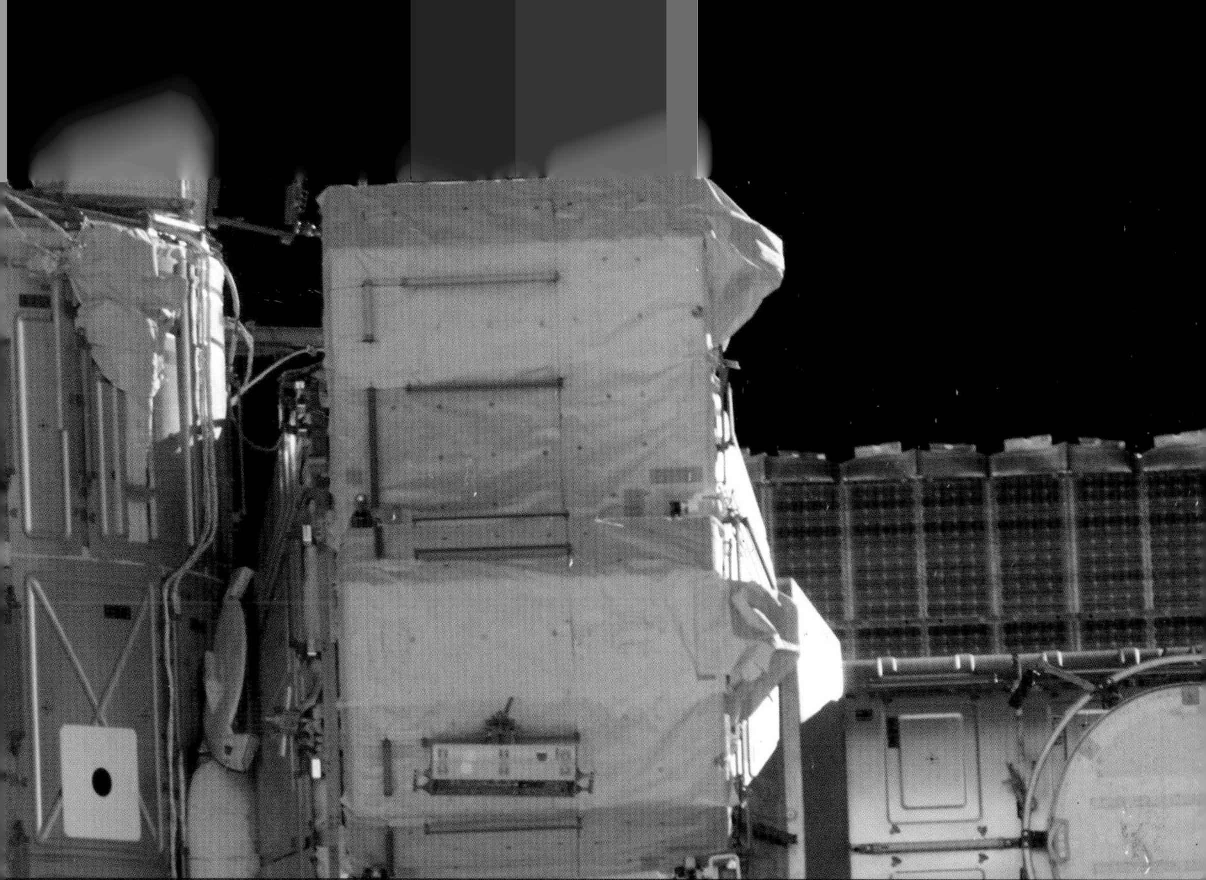

The International Space Station (ISS) is a great international, technological, and political achievement. It is the latest step in humankind's quest to explore and live in space. The results of the research done on the ISS may be applied in various areas of science, enable us to improve life on this planet, and give us the experience and increased understanding that can eventually equip us to journey to other worlds.

This book is designed to provide a broad overview of the Station's complex configuration, design, and component systems, as well as the sophisticated procedures required in the Station's construction and operation.

The ISS is in orbit today, operating with a crew of three. Its assembly will continue through 2010. As the ISS grows, its capabilities will increase, thus requiring a larger crew. Currently, 16 countries are involved in this venture.

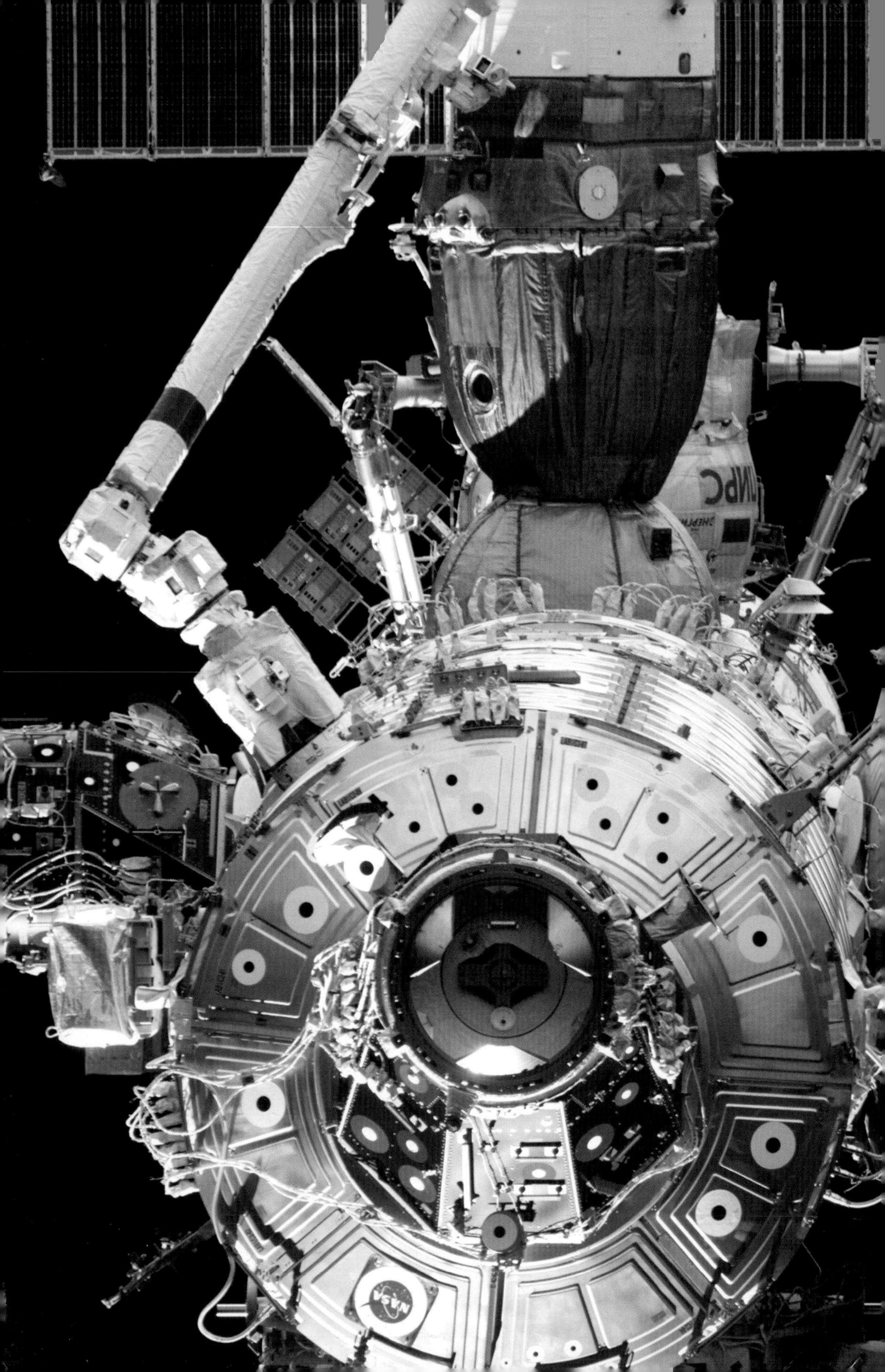
NASA

contents

Assembly Stages 9

Defines the characteristics of the ISS today, the design as it will be upon completion, and all of the assembly stages that have changed its appearance from the beginning through to Assembly Complete.

Elements 21

Describes the characteristics of each principal module of the ISS.

Transportation/Logistics 37

Describes the launch vehicles and carriers required to transport the components, crews, and consumables that support the ISS throughout its mission.

Systems 47

Provides an overview of each functional grouping of ISS hardware and the truss assembly, which serves as the structural backbone of the ISS.

ISS International Facilities and Operations 71

Details the international partners' principal locations, installations, and activities.

Missions 81

Displays a quick guide to the flights flown to the ISS and the crews responsible for its construction and operation.

Interesting Facts 91

Appendix 99

Contacts 103

Shown in the foreground, a telephoto view of the U.S. Lab. Clockwise from the left, the Pressurized Mating Adapter, the Space Station Remote Manipulator System, Soyuz, and Pirs. In the background, the U.S. Airlock.

The International Space Station (ISS) affords a unique opportunity to serve as an engineering test bed for flight systems and operations critical to NASA's exploration mission. U.S. research on the ISS will concentrate on the long-term effects of space travel on humans and engineering development activities in support of exploration. This research will help enable human crews to venture through the increasingly longer missions and greater distances necessary to visit Earth's planetary neighbors.

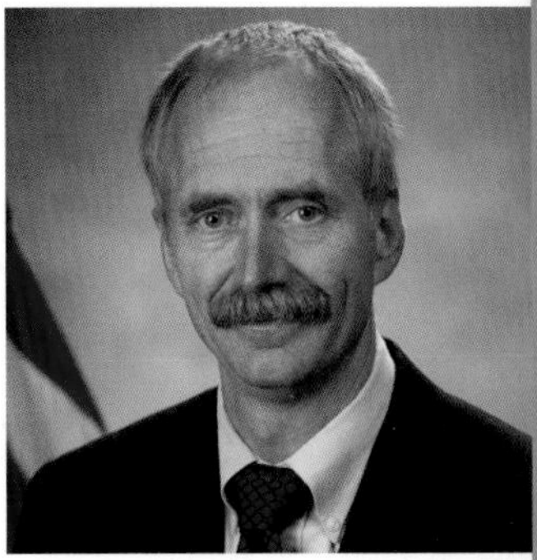

The National Aeronautics and Space Administration (NASA) looks forward to working with our partners on ISS research and engineering development and operations that will help open up new pathways for future exploration and discovery beyond low-Earth orbit.

—*William H. Gerstenmaier*
Associate Administrator
NASA Space Operations Mission Directorate

Telephoto close-up. Soyuz to left. Space Station Remote Manipulator System extends over Pressurized Mating Adapter 3. Functional Cargo Block in foreground.

assembly stages

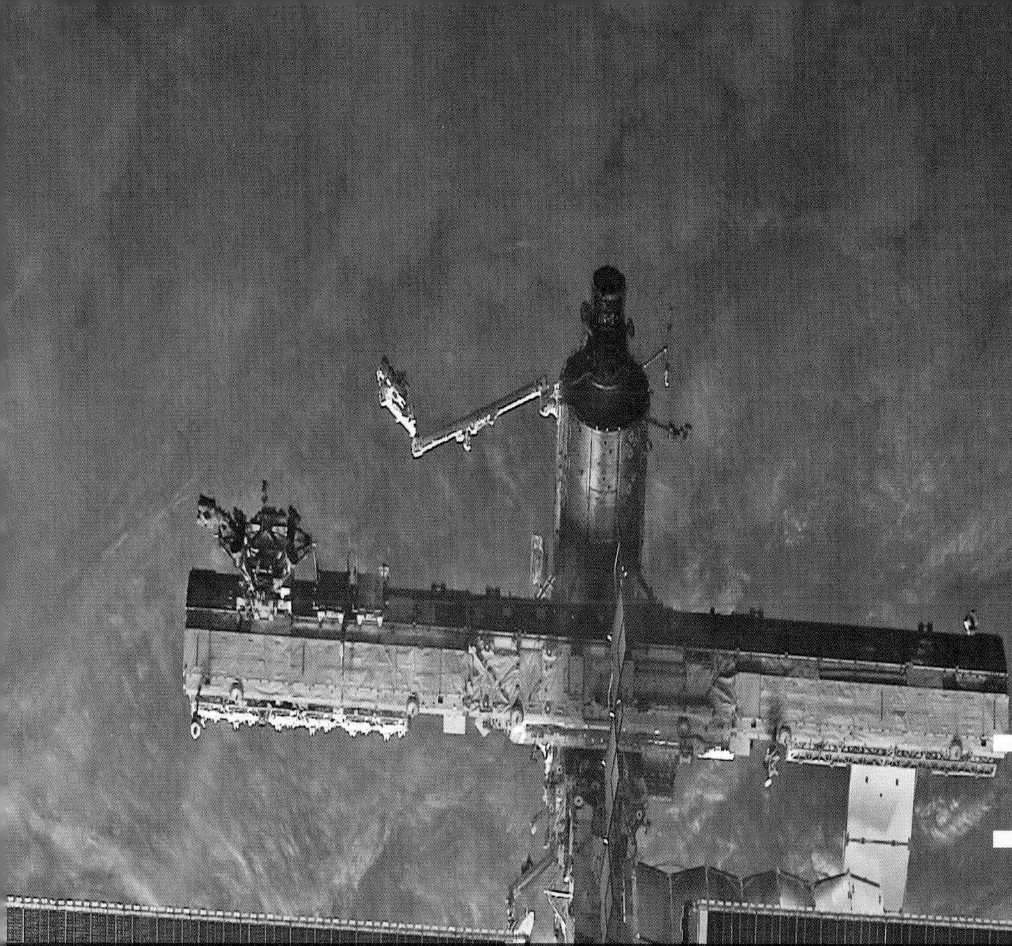

As of mid-2006, the International Space Station (ISS) has been continuously crewed for more than 5 years and is about 50 percent complete with approximately 180 metric tonnes (198 tons) of mass on orbit. There are 16 elements in orbit today, 9 elements ready for launch at Kennedy Space Center in Florida, and 6 elements in process at international partner sites. When the assembly is complete,the ISS will be composed of about 420,000 kilograms (925,000 pounds) of hardware brought to orbit in about 40 separate launches over the course of more than a decade. To date,there have been over 50 flights to the ISS, including flights for assembly, crew rotation, and logistical support.

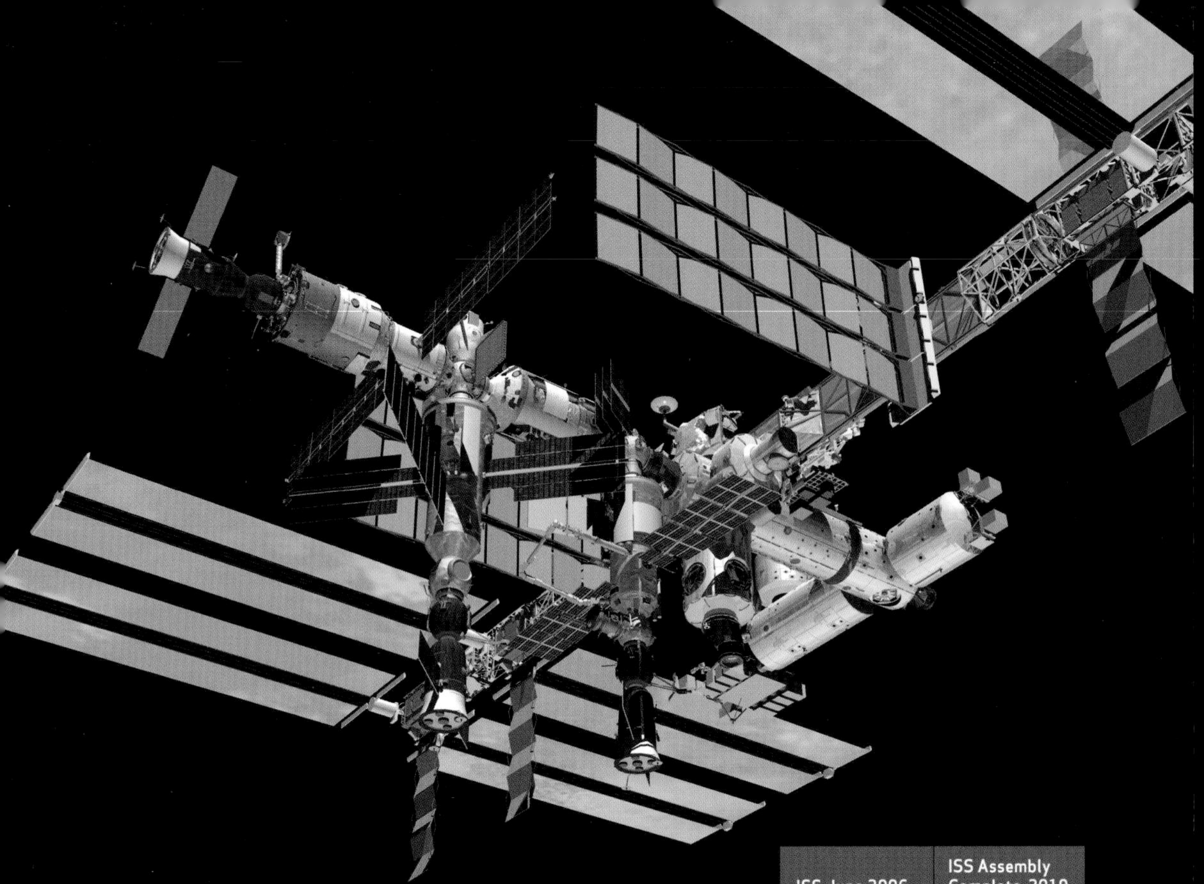

	ISS, June 2006	ISS Assembly Complete, 2010
Length	52 m (171 ft)	74 m (243 ft)
Width	73 m (240 ft)	110 m (361 ft)
Mass	186,000 kg (410,000 lb)	419,600 kg (925,000 lb)
Pressurized volume	449 m^3 (15,860 ft^3)	935 m^3 (33,023 ft^3)
Array surface area	892 m^2 (9,600 ft^2)	2,500 m^2 (27,000 ft^2)
Power	26 kW	110 kW

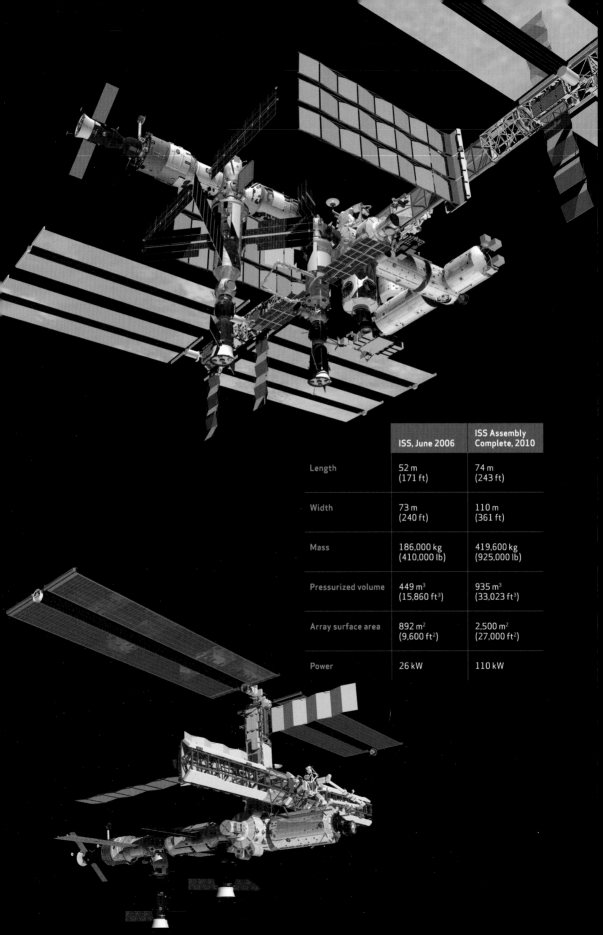

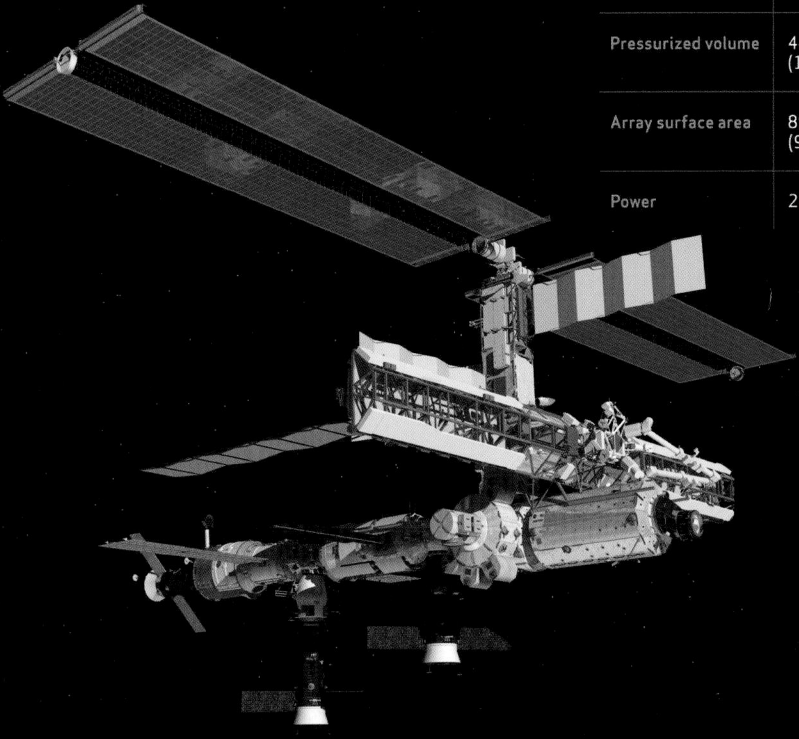

Principal Stages in Construction

The ISS, at Assembly Complete, is to be the largest humanmade object ever to orbit Earth. The ISS is to have a pressurized volume of 935 m^3 (33,023 ft^3), a mass of 419,600 kg (925,000 lb), maximum power output of 110 kW, with a payload long-term average power allocation of 30 kW, a structure that measures 110 m (361 ft) (across arrays) by 74 m (243 ft) (module length), an orbital altitude of 370–460 km (230–286 mi), an orbital inclination of 51.6º, and a crew of six.

Building and sustaining the ISS requires 80 flights over a 12-year period. As of 2006, 21 flights have been flown in support of ISS assembly. As many as another 17 Shuttle missions and 2 Russian launches are currently planned to complete the assembly. Currently, logistics is supported by the Space Shuttle, Progress, and Soyuz.

Future logistics/resupply missions will also be provided by the European Automated Transfer Vehicle (ATV) and Japan's H-II Transfer Vehicle (HTV). The U.S. Crew Exploration Vehicle (CEV) and commercial systems will support ISS logistics in the future.

STAGE/DATE	ELEMENT ADDED	LAUNCH VEHICLE	
1A/R Nov. 1998	 Functional Cargo Block (FGB).	Proton	
2A Dec. 1998	 Node 1, Pressurized Mating Adapter (PMA) 1, 2.	Space Shuttle STS-88	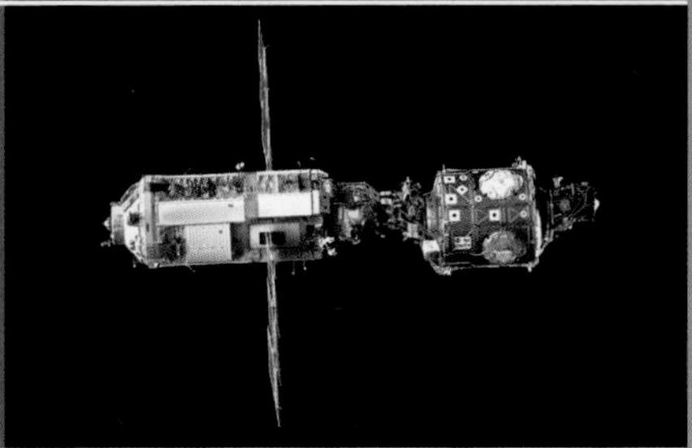
1R July 2000	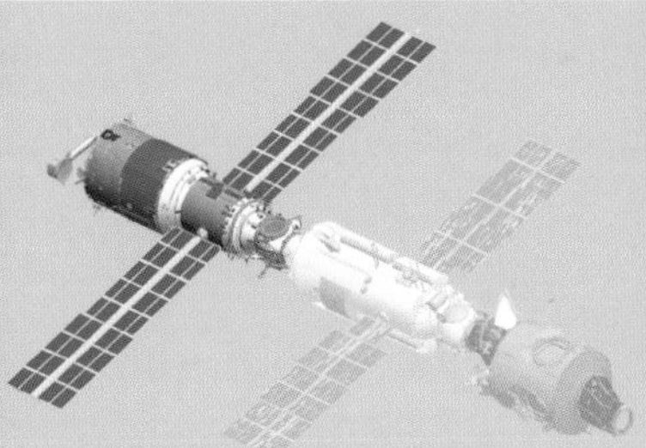 Service Module (SM).	Proton	
3A Oct. 2000	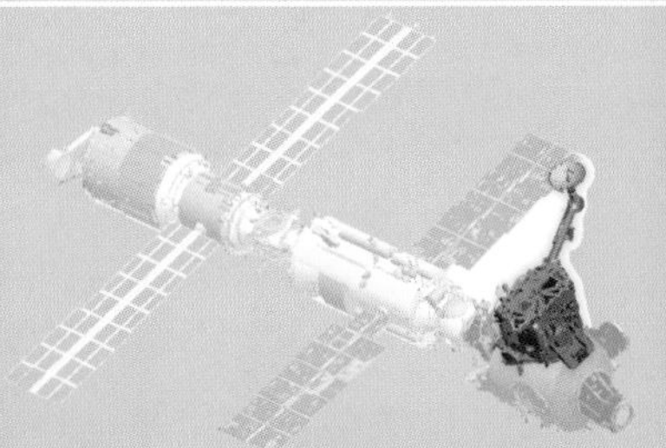 Zenith 1 (Z1) Truss, PMA 3.	Space Shuttle STS-92	

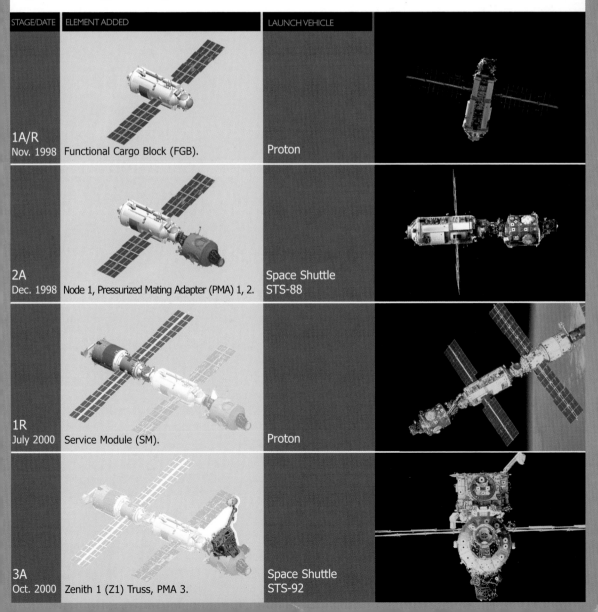

STAGE/DATE	ELEMENT ADDED	LAUNCH VEHICLE	
4A Dec. 2000	 Port 6 (P6) Truss.	Space Shuttle STS-97	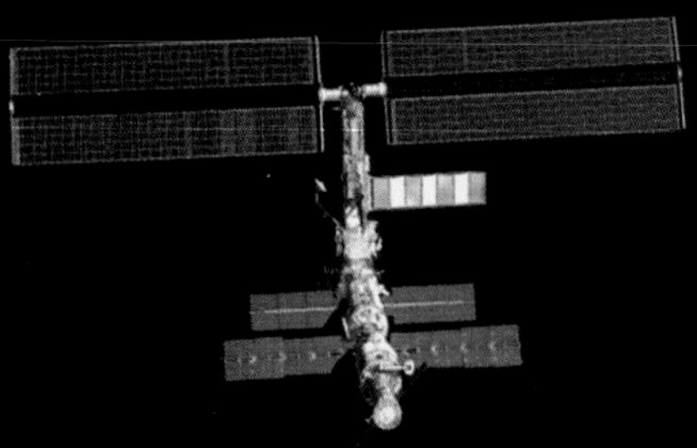
5A Feb. 2001	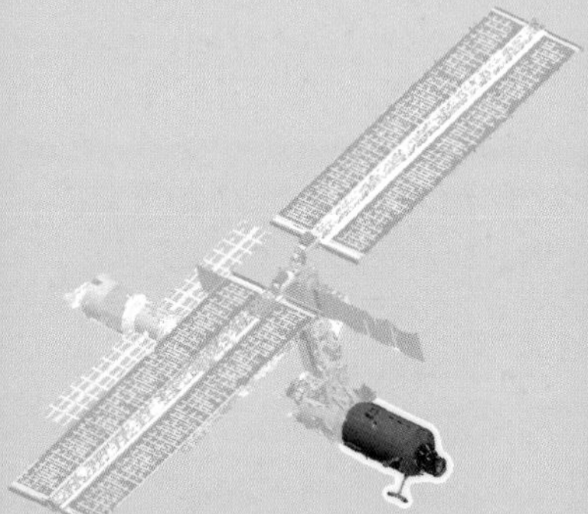 U.S. Lab.	Space Shuttle STS-98	
6A Apr. 2001	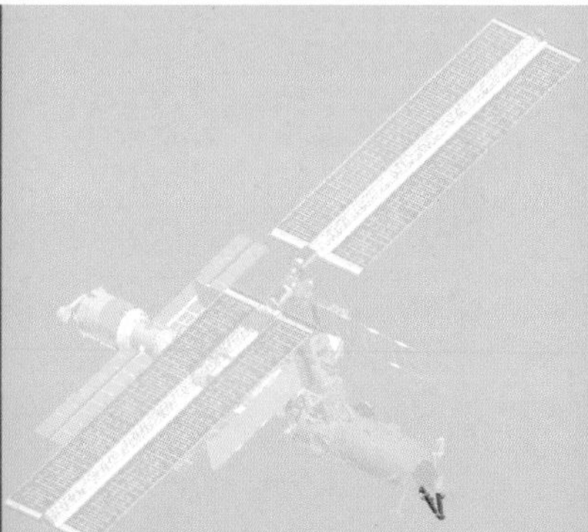 Space Station Remote Manipulator System (SSRMS).	Space Shuttle STS-100	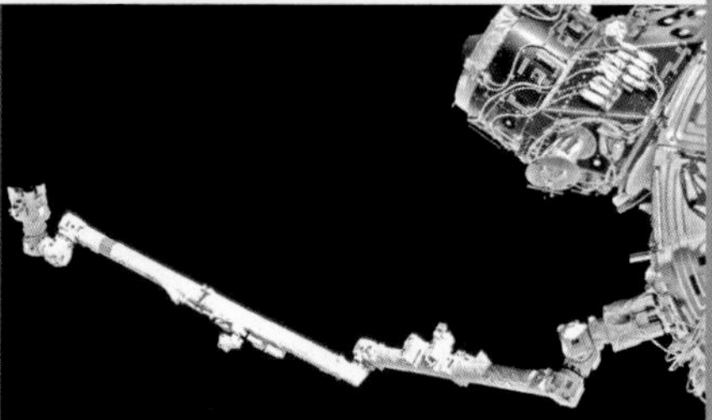
7A July 2001	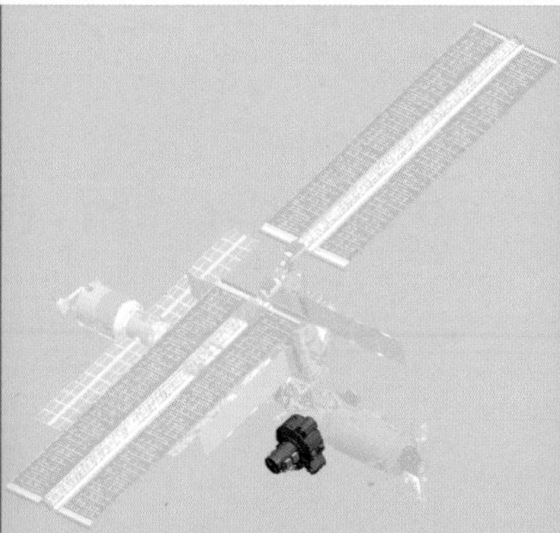 U.S. Airlock.	Space Shuttle STS-104	

STAGE/DATE	ELEMENT ADDED	LAUNCH VEHICLE	
4R Sep. 2001	 Russian Docking Compartment (DC) and Airlock.	Soyuz	
8A Apr. 2002	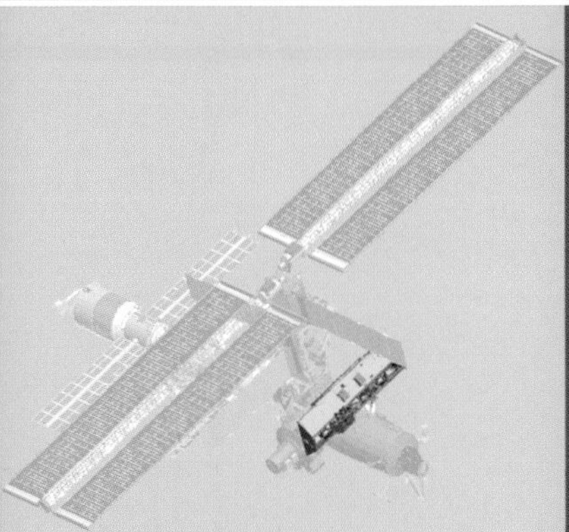 Starboard Zero (S0) Truss.	Space Shuttle STS-110	
9A Oct. 2002	 S1 Truss.	Space Shuttle STS-112	
11A Nov. 2002	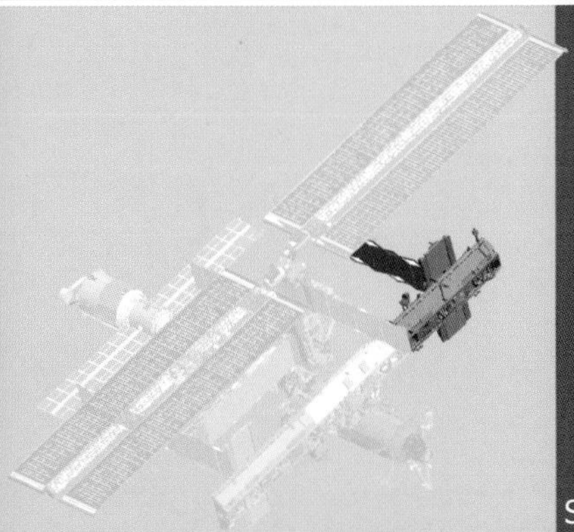 P1 Truss.	Space Shuttle STS-113	

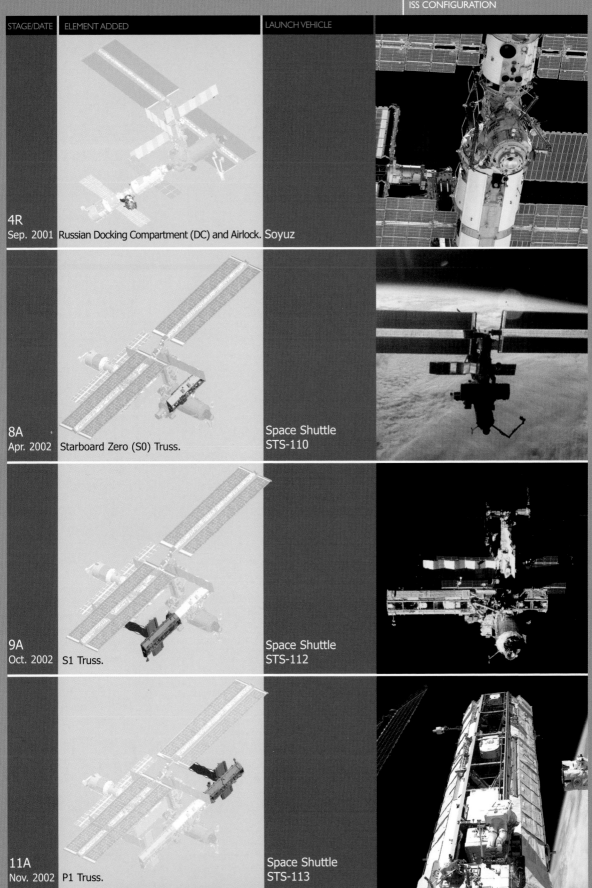

STAGE/DATE	ELEMENT ADDED	LAUNCH VEHICLE	
12A Aug. 2006	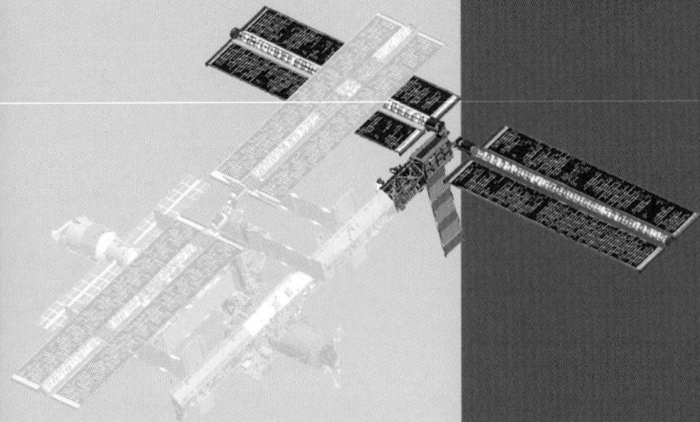 P3/P4 Truss.	Space Shuttle STS-115	
12A.1	 P5 Truss, retracting P6 arrays.	Space Shuttle STS-116	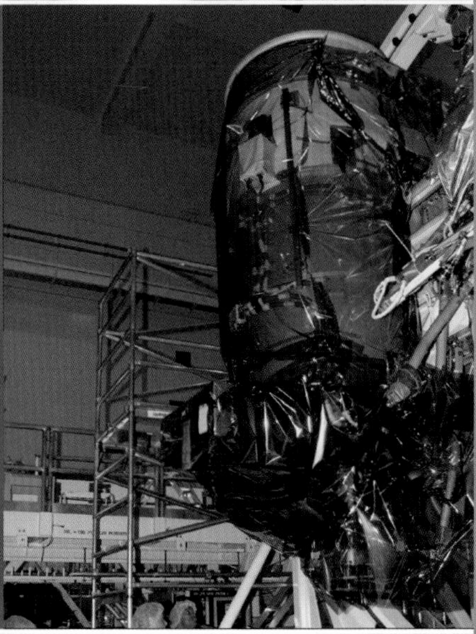
13A	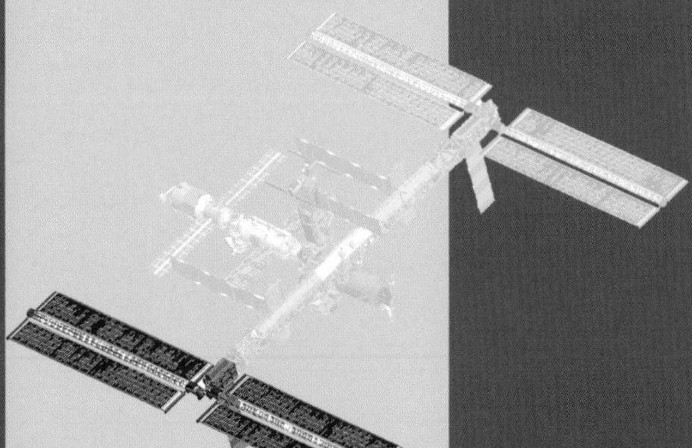 S3/S4 Truss.	Space Shuttle STS-117	

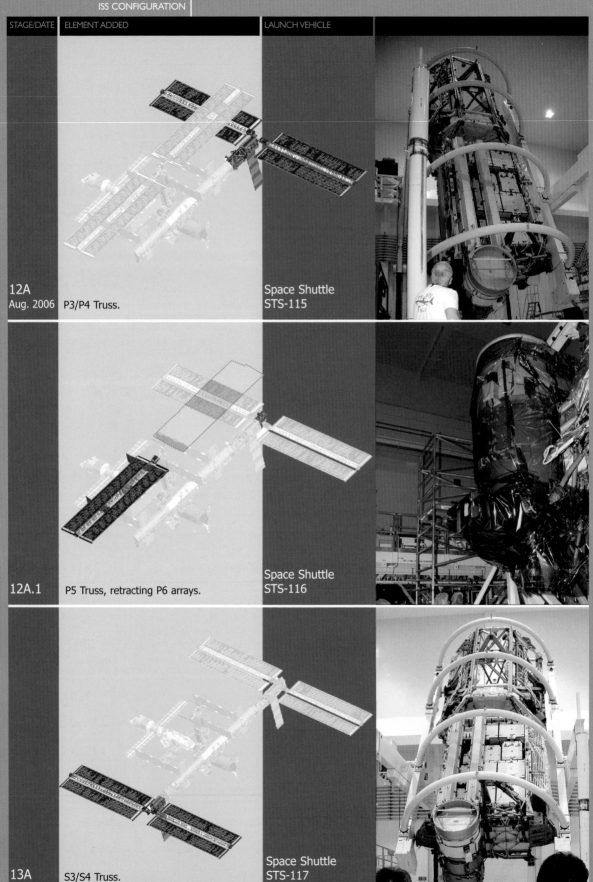

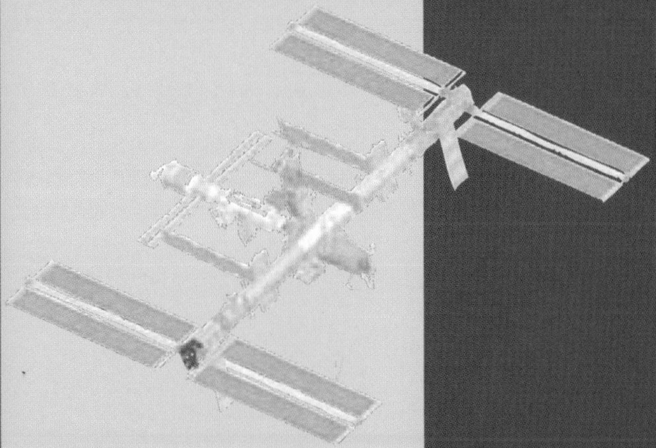

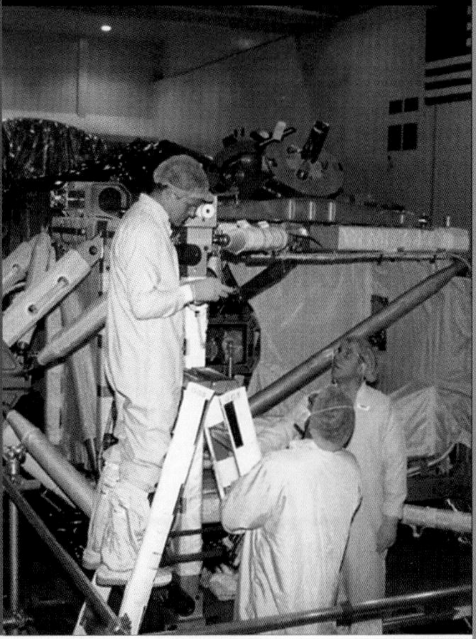

STAGE/DATE	ELEMENT ADDED	LAUNCH VEHICLE
13A.1	S5 Truss.	Space Shuttle STS-118
10A	S5 Truss.	Space Shuttle STS-120
1E	ESA Columbus Module.	Space Shuttle STS-122

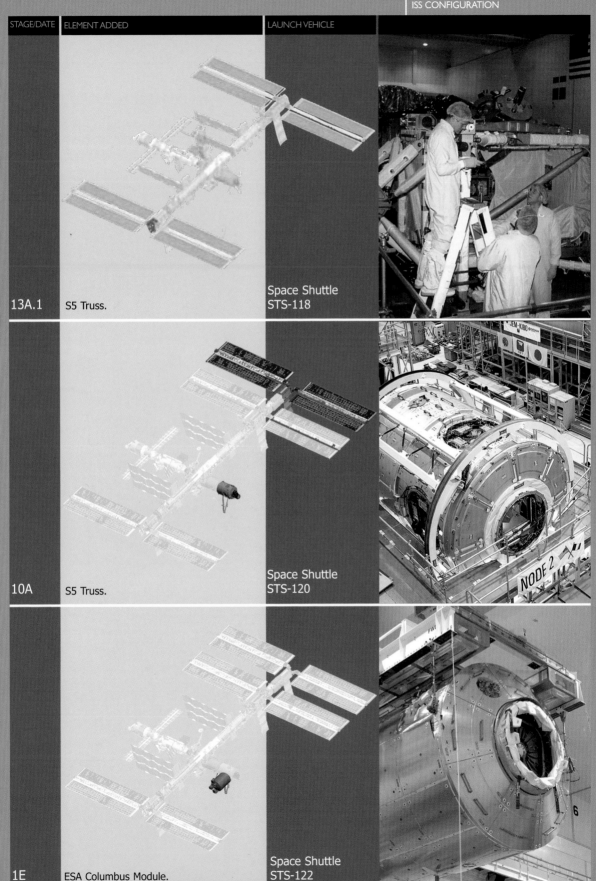

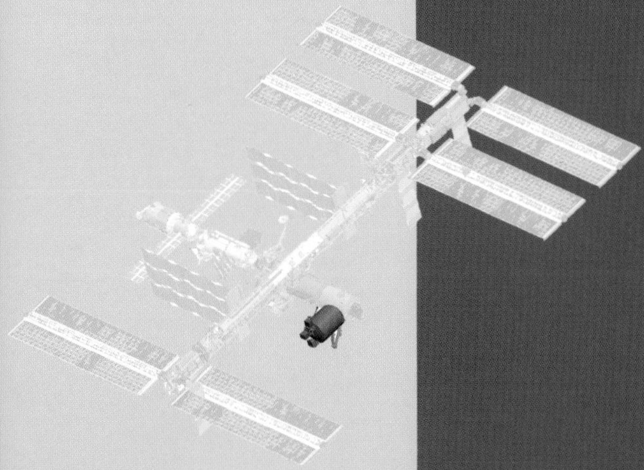

STAGE/DATE	ELEMENT ADDED	LAUNCH VEHICLE	
1J/A	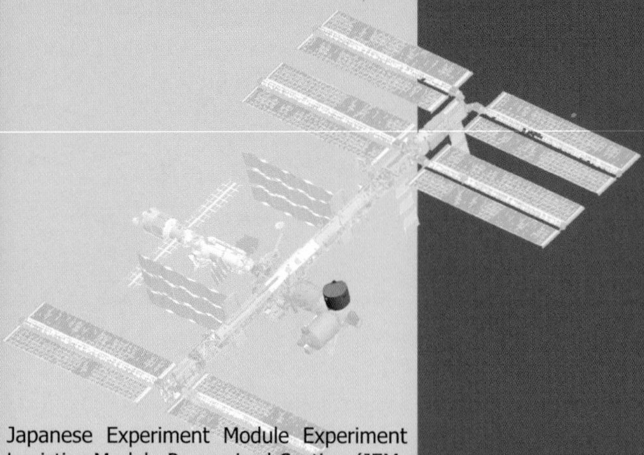 Japanese Experiment Module Experiment Logistics Module Pressurized Section (JEM-ELM-PS), and Canadian Special Purpose Dexterous Manipulator (Dextre).	Space Shuttle	
1J	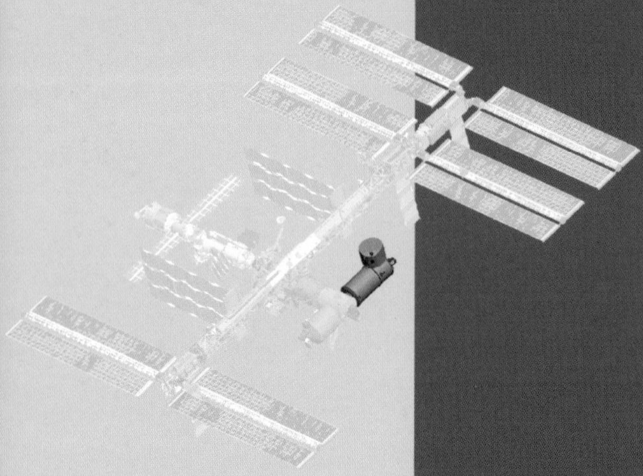 JEM Pressurized Module (PM).	Space Shuttle	
15A	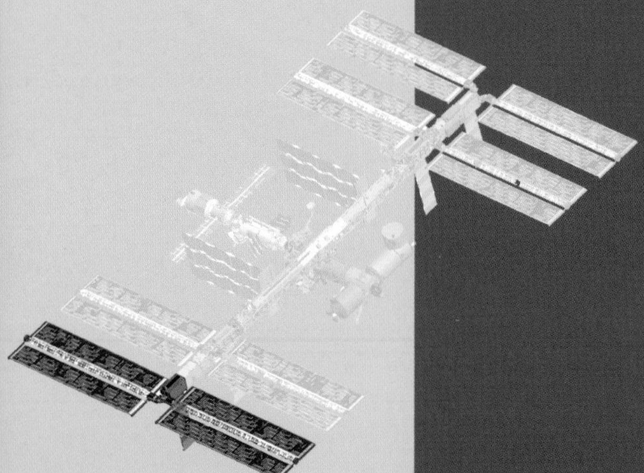 S6 Truss.	Space Shuttle	

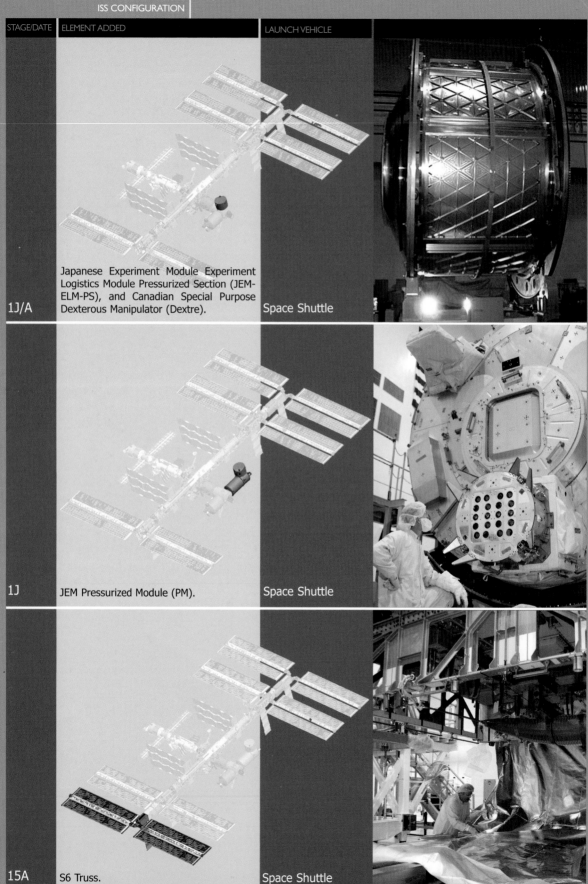

STAGE/DATE	ELEMENT ADDED	LAUNCH VEHICLE	
2J/A	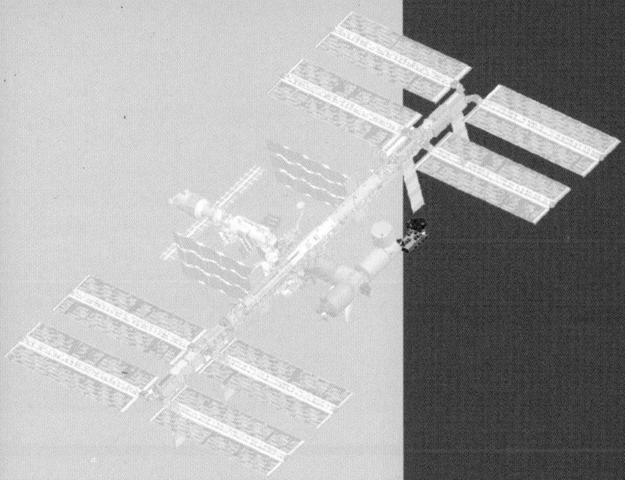 JEM-ELM Exposed Section (ES), JEM-Exposed Facility (JEM-EF).	Space Shuttle	
3R	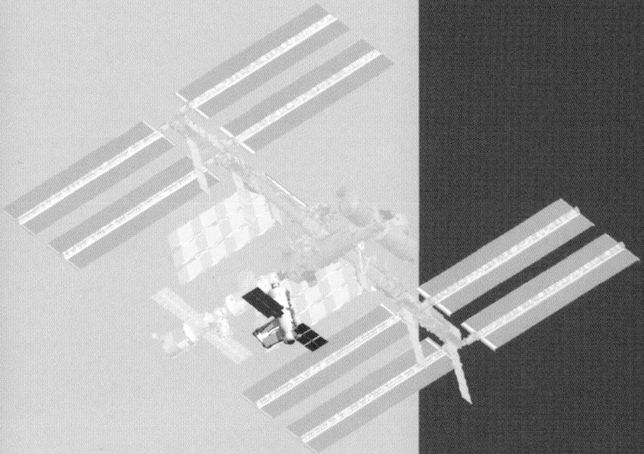 Russian Multi-Purpose Laboratory Module.	Proton	
20A	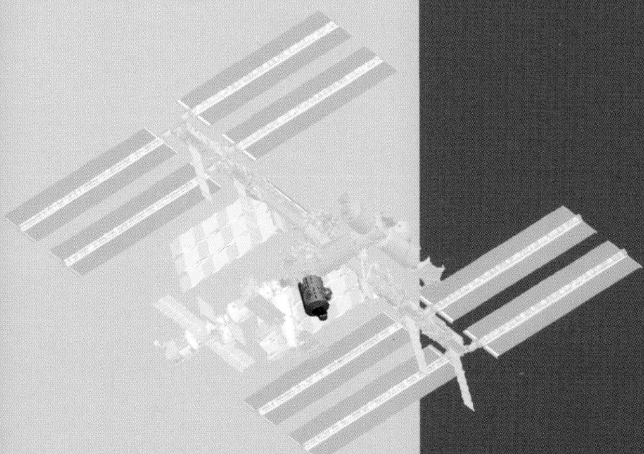 Node 3 and Cupola.	Space Shuttle	

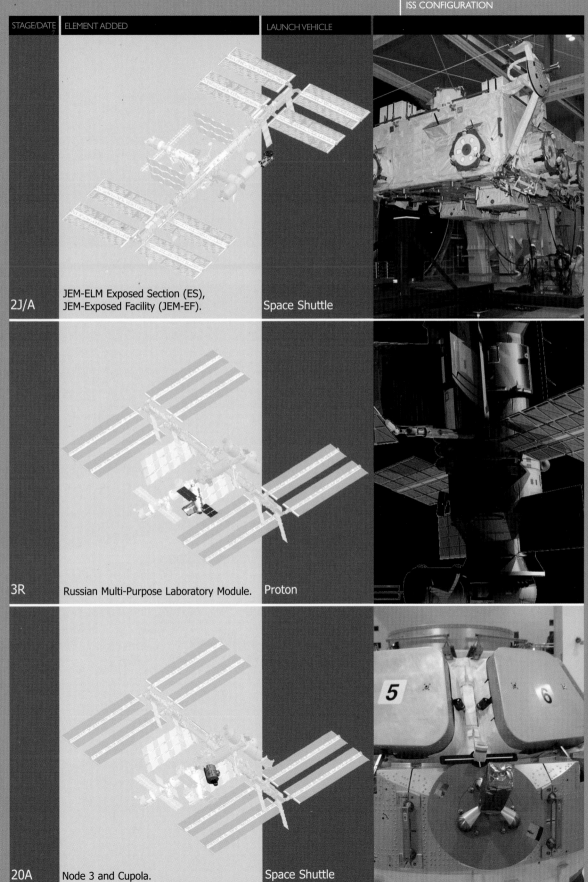

STAGE/DATE	ELEMENT ADDED	LAUNCH VEHICLE	
9R	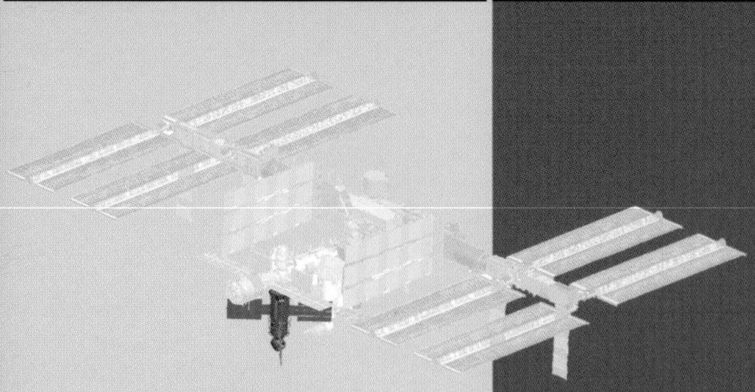 Russian Research Module.	Proton	

A=U.S. Assembly J=Japanese Assembly E=European Assembly R=Russian Assembly

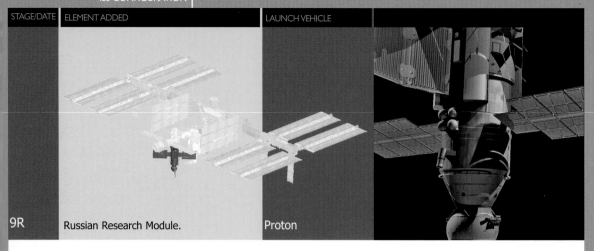

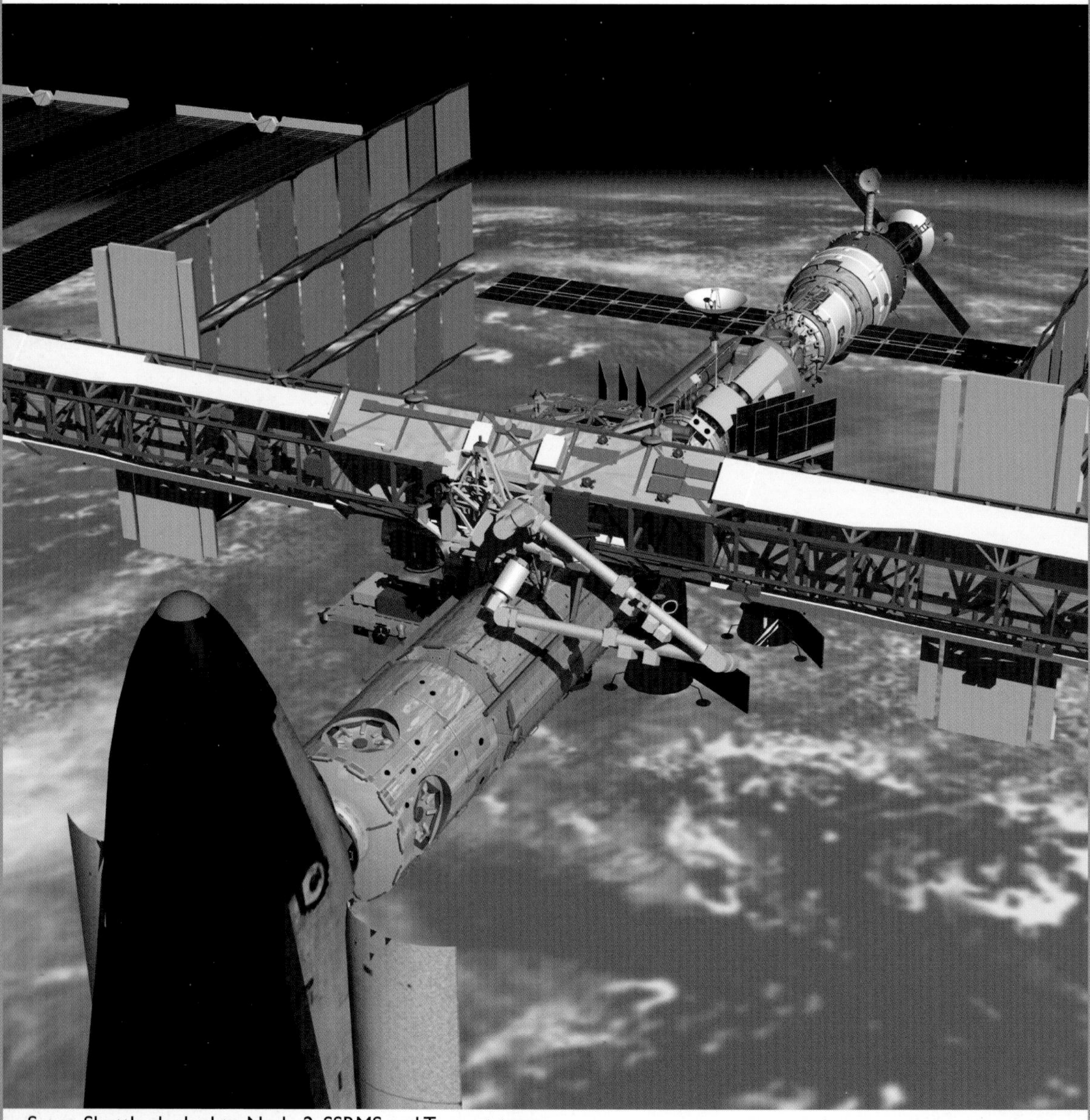

Space Shuttle docked to Node 2. SSRMS and Truss at top.

Current ISS on-orbit elements

MODULE	LENGTH	MASS	LAUNCHED	LAUNCH VEHICLE
FGB (Zarya)	12.8 m (42 ft)	25,000 kg (55,000 lb)	11/20/98	Proton
Node 1 (Unity)/PMA 1 & 2	10.4 m (34 ft)	14,900 kg (33,000 lb)	12/04/98	STS-88
Service Module (Zvezda)	13.1 m (43 ft)	24,600 kg (54,200 lb)	7/12/00	Proton
Z1 Truss/PMA 3	4.6 m (15 ft)	8,755 kg (19,227 lb)/ 1,168 kg (2,575 lb)	10/11/00	STS-92
P6 Truss	18.3 m (60 ft) 73.2 m (240 ft) across extended solar array	14,550 kg (32,100 lb)	11/30/00	STS-97
U.S. Lab (Destiny)	8.5 m (28 ft)	24,100 kg (53,100 lb)	02/07/01	STS-98
SSRMS (Canadarm 2)	17.7 m (58 ft)	1,502 kg (3,311 lb)	04/19/01	STS-100
U.S. Airlock (Quest)	4.6 m (15 ft)	9,920 kg (21,900 lb)	07/12/01	STS-104
Docking Compartment/ Airlock (Pirs)	4.9 m (16 ft)	3,838 kg (8,461 lb)	09/15/01	Soyuz
S0 Truss/Mobile Transporter	13.4 m (44 ft)	12,100 kg (26,700 lb)	04/08/02	STS-110
Mobile Base System	5.8 m (19 ft)	1,450 kg (3,200 lb)	06/05/02	STS-111
S1 Truss	13.7 m (45 ft)	12,300 kg (27,100 lb)	10/07/02	STS-112
P1 Truss	13.7 m (45 ft)	12,300 kg (27,100 lb)	11/23/02	STS-113
Soyuz (typical)	7 m (22.9 ft)	7,167 kg (15,800 lb)	N/A	Soyuz
Progress (typical)	7.3 m (24 ft)	1,750 kg (15,800 lb)	N/A	Soyuz

Current and Future totals

	LENGTH	WIDTH	VOLUME	MASS
June 2006	52 m (171 ft) with Progress	73 m (240 ft) across array	449 m³ (15,860 ft³)	186,000 kg (410,000 lb) 186 t (205 tons)
Assembly Complete	74 m (243 ft) with ESA ATV	108.5 m (356 ft) arrays extended	935 m³ (33,023 ft³)	419,600 kg (925,000 lb) 457 t (420 tons)

ISS Assembly Sequence

The table below shows the plan for completion. Assembly and logistics flights are plotted as a function of time and percent of total mass.

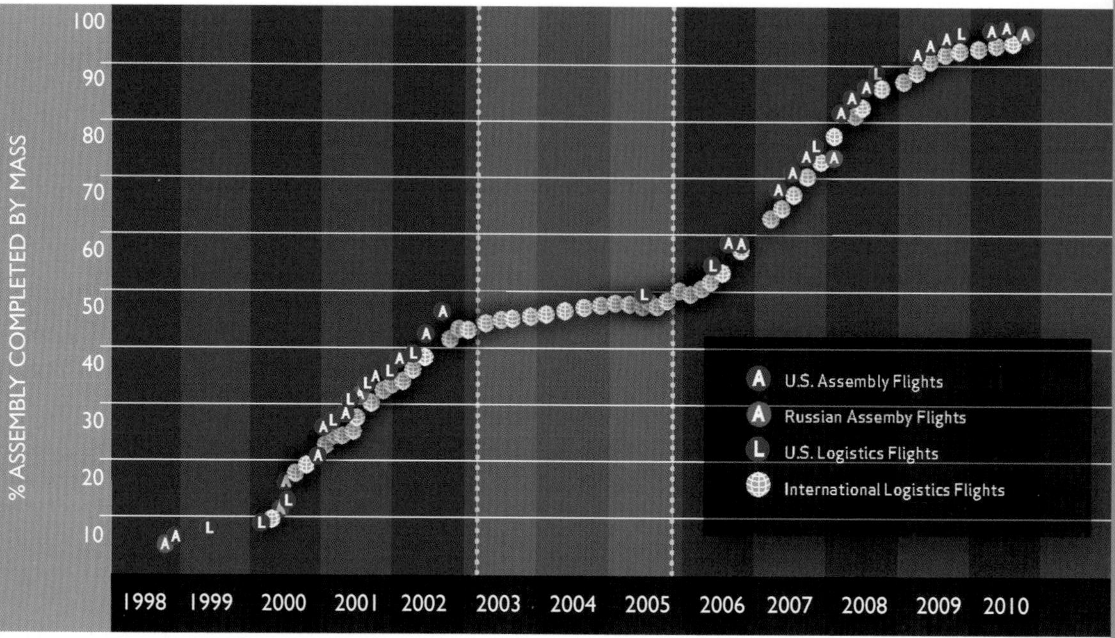

Important Dates

Nov. 20, 1998	First element launched (FGB)
Dec. 4, 1998	Shuttle mission carried first U.S. component, Node 1 (Unity)
July 12, 2000	Early living quarters launched by Russians, Service Module (Zvezda)
Nov. 2, 2000	Start of permanent human presence on the ISS (Expedition 1)
Nov. 2000	First set of U.S. arrays made the ISS the most powerful spacecraft ever
Feb. 2001	U.S. laboratory Destiny delivered (provided command and control and an experiment platform)
Apr. 2001	Canadian robotic arm extended the "reach" of the Station for assembly
July 2001	U.S. airlock Quest arrived, allowing U.S. spacewalks without the Shuttle
Apr. 2002	S0 Truss (central truss segment); Mobile Transporter launched
June 2002	Mobile Base System (platform on which SSRMS can attach for translation across truss) installed
Sept. 2002	S1 Truss installed
Nov. 2002	P1 Truss installed
July 2005	Space Shuttle Return to Flight (STS-114)—a logistics mission
	Six-person crew
	Assembly Complete

elements

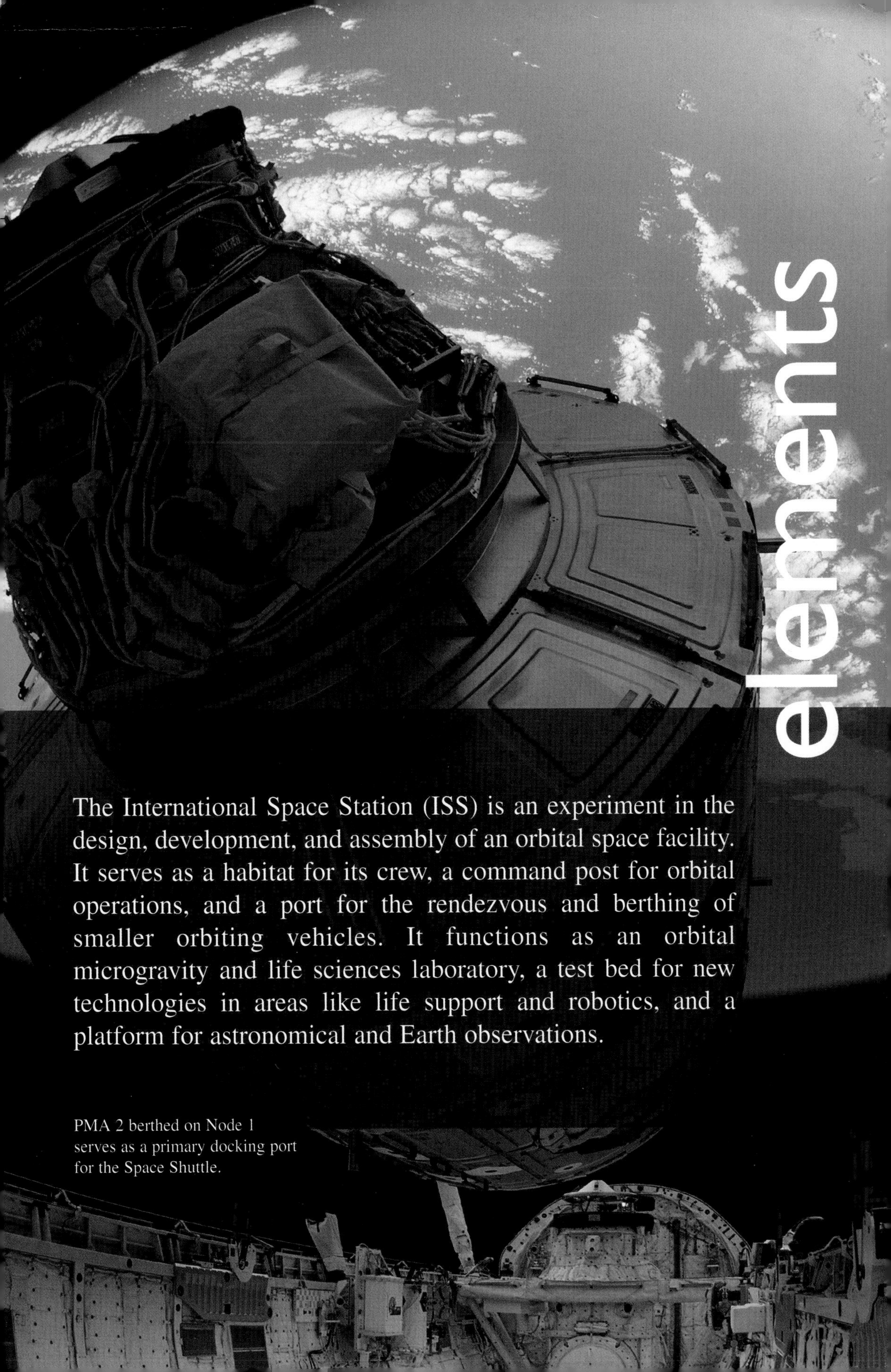

The International Space Station (ISS) is an experiment in the design, development, and assembly of an orbital space facility. It serves as a habitat for its crew, a command post for orbital operations, and a port for the rendezvous and berthing of smaller orbiting vehicles. It functions as an orbital microgravity and life sciences laboratory, a test bed for new technologies in areas like life support and robotics, and a platform for astronomical and Earth observations.

PMA 2 berthed on Node 1 serves as a primary docking port for the Space Shuttle.

The U.S. Lab Module Destiny provides research and habitation accommodations. Node 2 is to the left; the truss is mounted atop the U.S. Lab; Node 1, Unity, is to the right; Node 3 and the Cupola are below and to the right.

Architecture Design Evolution

Why does the ISS look the way it does ?
The design evolved over more than a decade. The modularity and size of the U.S., Japanese, and European elements were dictated by the use of the Space Shuttle as the primary launch vehicle and by the requirement to make system components maintainable and replaceable over a lifetime of many years.

When the Russians joined the program in 1993, their architecture was based largely on the Mir and Salyut stations they had built earlier. Russian space vehicle design philosophy has always emphasized automated operation and remote control.

The design of the interior of the U.S., European, and Japanese elements was dictated by four specific principles: modularity, maintainability, reconfigurability, and accessibility. Interior modular hardware racks and utilities could be replaced as needs or age dictated. Racks could be swung away from the pressure hull of the module in case a meteoritic puncture necessitated a repair. Crew preferences dictated that module interiors be arranged with distinct floors, ceilings, and walls.

Module Design and Layout

1979—Modules with connecting tunnels.

1982—Common modules.

1986—Habitation Module, Laboratory Module (Hab, Lab), spherical Nodes, and tunnels.

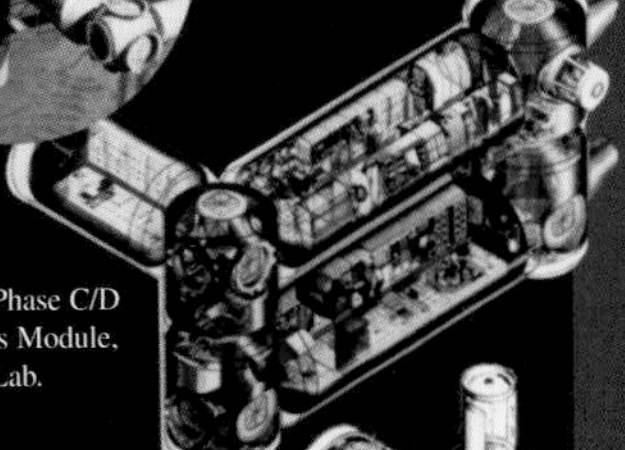

1988—Boeing Phase C/D Nodes, Logistics Module, and 45-ft Hab, Lab.

1992—Freedom, Nodes, Airlock, Logistics Module, and 27-ft Hab, Lab.

Module Architecture Early Concepts

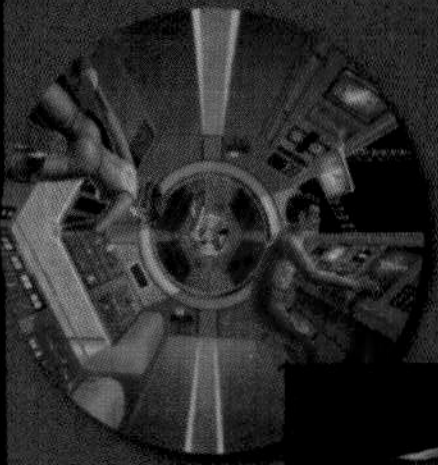

1980—Horizontal layout.

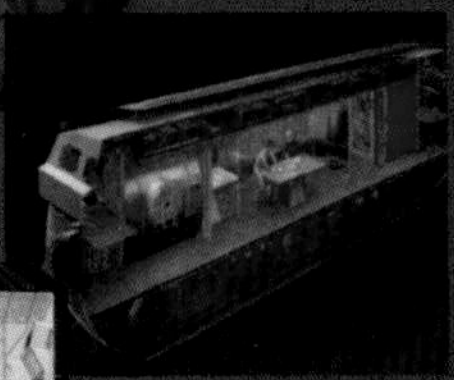

1980—Horizontal layout.

1980—Vertical layout.

1986—Central core.

1986—Central beam.

Module Architecture Racks with Four Structural Standoffs

Loft concept.

Modular outfitting. Standoff

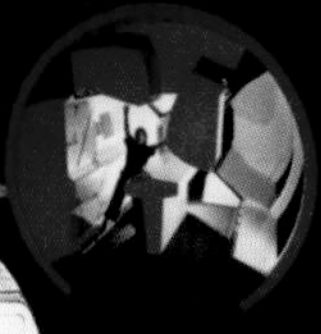

Standard racks (2 sizes).

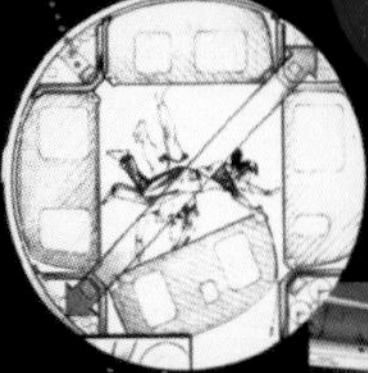

Standard rack (1 size).

Access to module pressure shell.

Access to utility runs in standoffs.

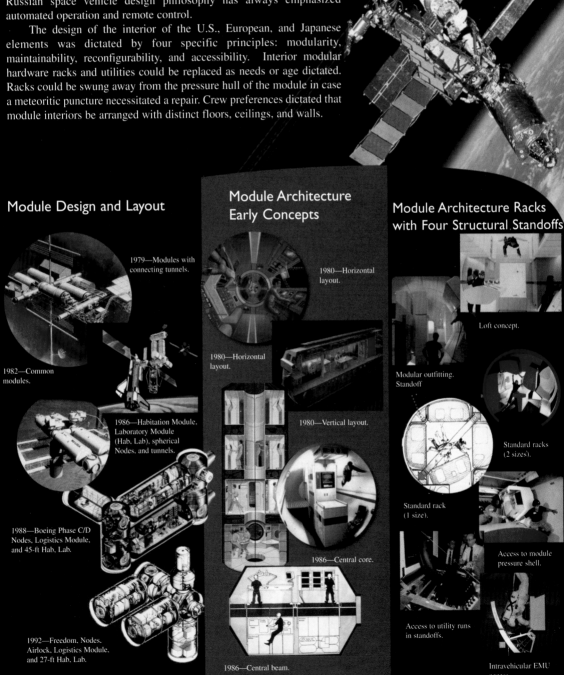

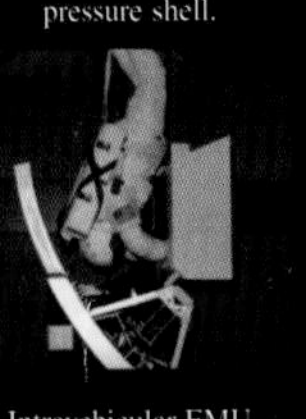

Intravehicular EMU access.

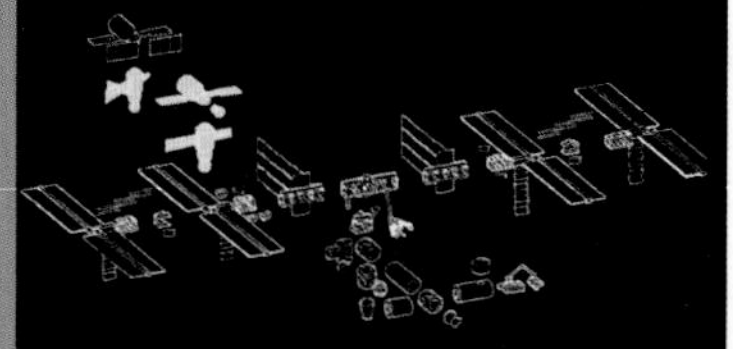

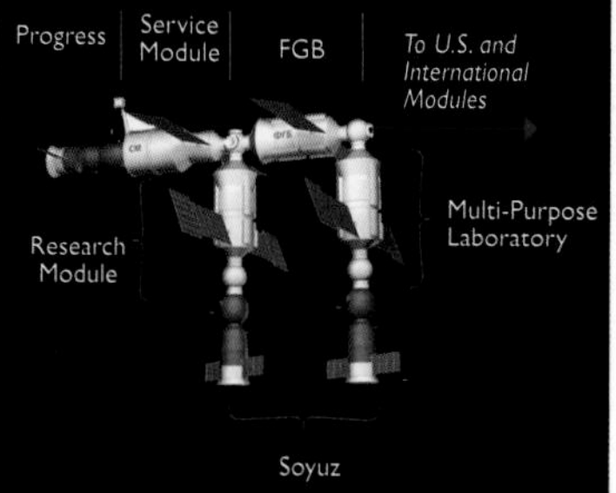

Functional Cargo Block (FGB)

Zarya (Sunrise) and Russian Research Modules

NASA/Khrunichev Production Center

The FGB was the first element of the International Space Station, built in Russia under a U.S. contract. During the early stages of ISS assembly, the FGB was self-contained, providing power, communications, and attitude control functions. The FGB module is now used primarily for storage and propulsion. The FGB was based on the modules of Mir. The Russian Multipurpose Modules planned for the ISS will be based on the FGB-2, a spare developed as a backup to the FGB. The Russian Research Module may be based on the FGB design.

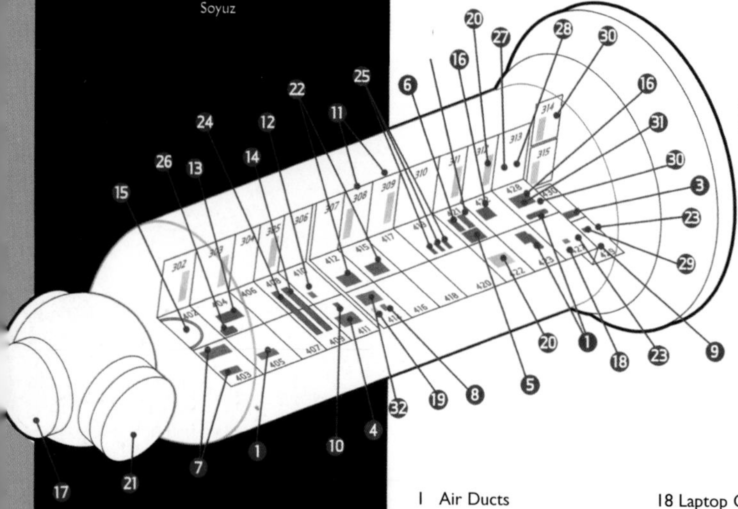

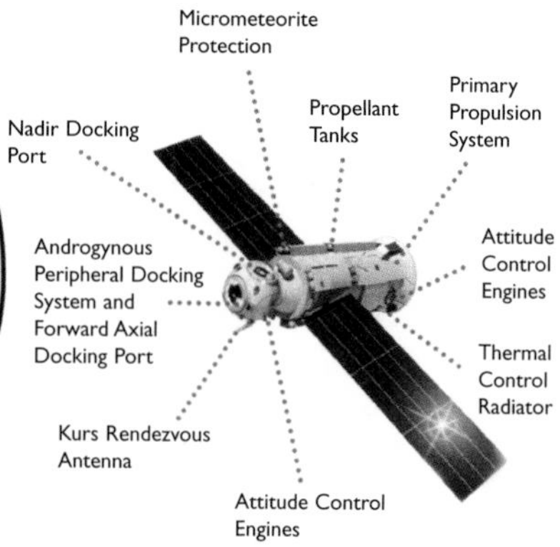

Length	12.990 m (42.6 ft)
Maximum diameter	4.1 m (13.5 ft)
Mass	24,968 kg (55,045 lb)
Pressurized volume	71.5 m³ (2,525 ft³)
Solar array span	24.4 m (80 ft)
Array surface area	28 m² (301 ft²)
Power supply (avg.)	3 kW
Propellant mass	3,800 kg (8,377 lb)
Launch date	Nov. 20, 1998, on a Proton rocket

1 Air Ducts
2 Communications Panel
3 Caution and Warning Systems Panel
4 Contaminant Filters
5 Contingency Transfer (Water) Container Bag
6 Contingency Transfer (Water) Container Connections
7 Dust Collectors
8 Electrical Outlet
9 Flex Airduct Container
10 Fuse
11 Fuse Panels (behind close-outs)
12 Gas Analyzer
13 Gas Mask
14 Handrail
15 Hatch Protection
16 Instrument Containers
17 Docking Port to PMA
18 Laptop Outlets
19 Lighting Panel
20 Lights
21 Nadir Docking Port
22 Onboard Documentation
23 Onboard Network Receptacle Outlets
24 Pole and Hook
25 Portable Fans
26 Removable Fire Extinguisher
27 Power Outlet
28 Pressurized Valve Unit
29 Caution and Warning Panel
30 Smoke Detector
31 TV Outlet
32 Wipes/Filters

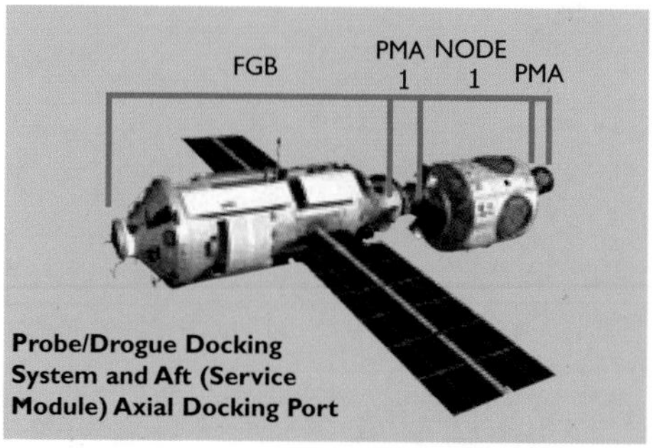

Probe/Drogue Docking System and Aft (Service Module) Axial Docking Port

Service Module (SM)

Zvezda (Star)

S.P. Korolev Rocket and Space Corporation Energia (RSC Energia)

The Service Module was the first fully Russian contribution to the ISS. The Module provided the Station's early living quarters, life-support system, electrical power distribution, data processing system, flight control system, and propulsion system. Its communications system still enables remote command capabilities from ground flight controllers. Although some of these systems were subsequently supplemented or replaced by later U.S. systems, the Service Module remains the structural and functional center of the Russian segment of the International Space Station.

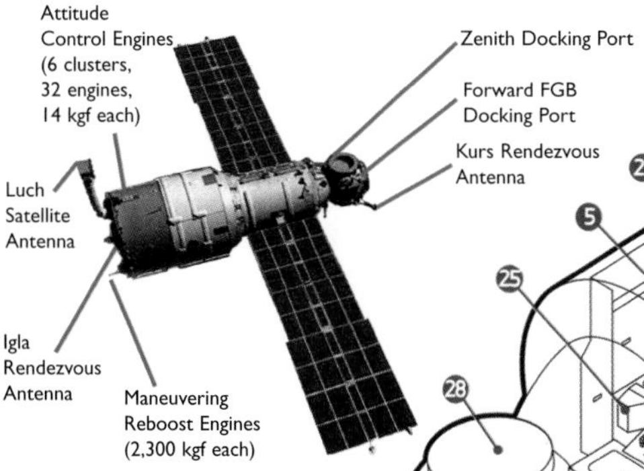

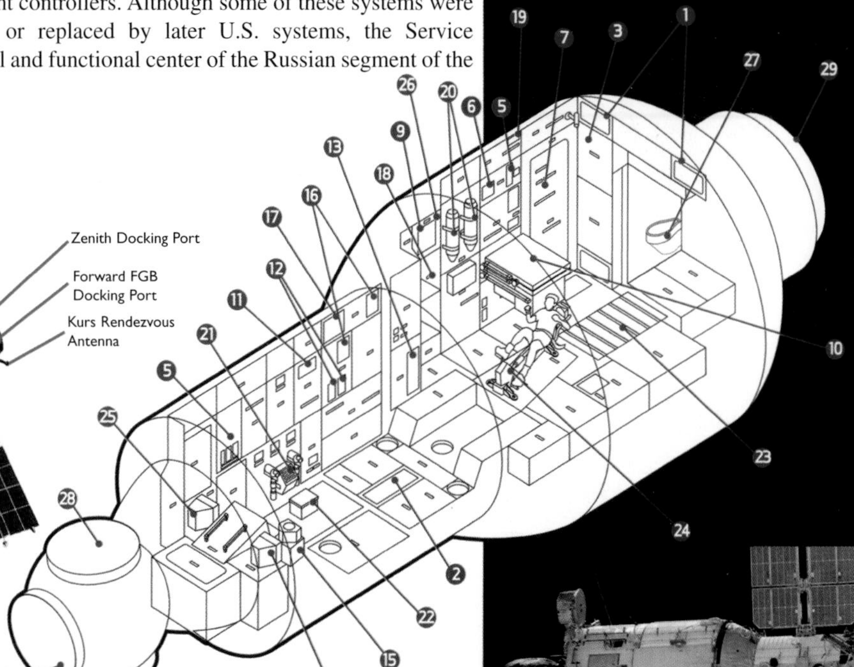

1 Airflow Vent
2 Body Mass Measurement Device
3 Camera
4 Caution and Warning Panel, Clock, and Monitors
5 Communications Panel
6 Condensate Water Processor
7 Crew Sleep Compartment
8 Forward Docking Port (to FGB)
9 Fuses
10 Galley Table
11 Integrated Control Panel
12 Lighting Control Panels
13 Maintenance Box
14 Nadir Docking Port
15 Navigation Sighting Station
16 Night-Lights
17 Power Distribution Panel
18 Recessed Cavity & Valve Panel
19 Smoke Detector
20 Solid Fuel Oxygen Generators (SFOG)
21 Toru Rendezvous Control Station
22 Toru Seat
23 Treadmill & Vibration Isolation System
24 Vela Ergometer
25 Ventilation Screen
26 Vozdukh Control Panel
27 Waste Management Compartment
28 Zenith Docking Port
29 Soyuz and Progress Docking Port

Length	13.1 m (43 ft)
Diameter	4.2 m (13.5 ft)
Wingspan	29.7 m (97.5 ft)
Weight	24,604 kg (54,242 lb)
Launch date	July 11, 2000, on a Proton rocket
Attitude control	32 engines
Orbital maneuvering	2 engines

The SM under construction at Khrunichev State Research and Production Space Center in Moscow.

Leroy Chiao exercises in the SM.

Pressurized Mating Adapters (PMAs)

NASA/Boeing

Three conical docking adapters, called Pressurized Mating Adapters, allow the docking systems used by the Space Shuttle and by Russian modules to attach to the Node's berthing mechanisms. PMA 1 links the U.S. and Russian segments. The other two adapters serve as docking ports for the Space Shuttle and will do the same for the Crew Exploration Vehicle (CEV) and later commercial vehicles.

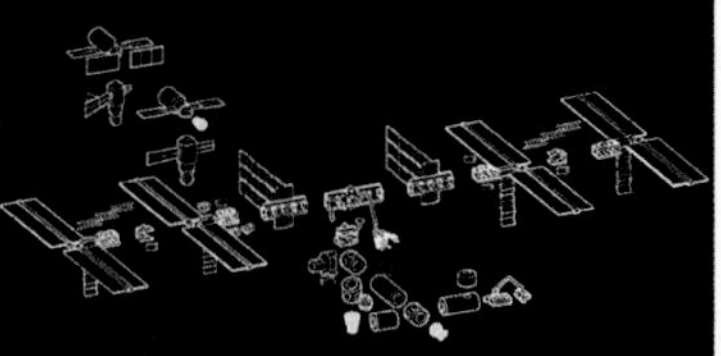

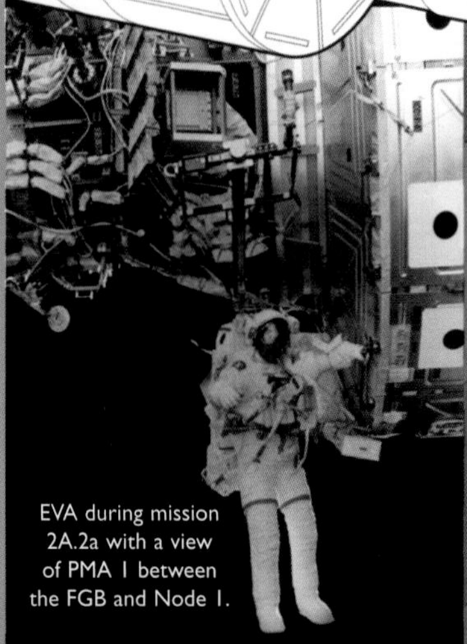

EVA during mission 2A.2a with a view of PMA 1 between the FGB and Node 1.

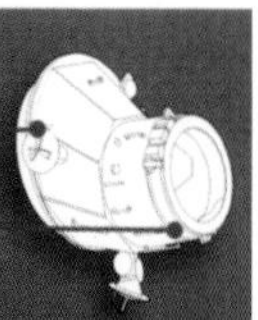

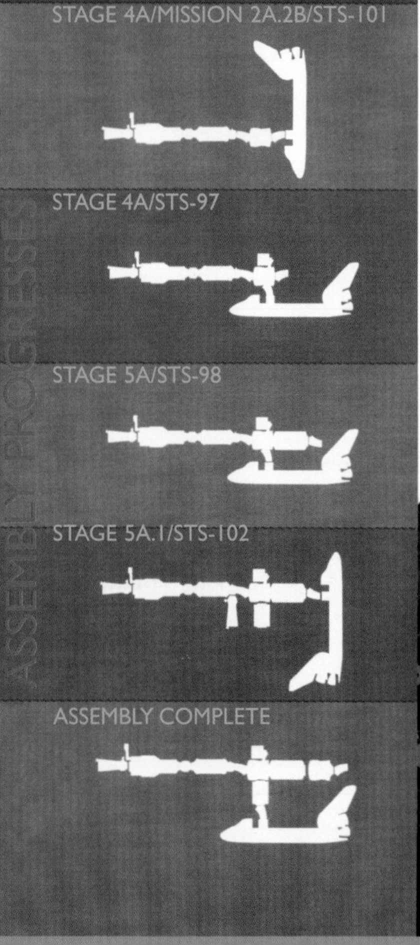

The PMA 1, 2, and 3 structures are identical and provide a pressurized interface between the U.S. and Russian ISS modules and between the U.S. modules and the Space Shuttle orbiter. The PMA structure is a truncated conical shell with a 28-inch axial offset in the diameters between the end rings.

PMA 1 is being attached to the Common Berthing Mechanism of Node 1.

PMA 2 on the forward berthing ring of Node 1.

Length	1.86 m (6.1 ft)
Width	1.9 m (6.25 ft) at wide end 1.37 m (4.5 ft) at narrow end
Mass of PMA 1 PMA 2 PMA 3	 1,589 kg (3,504 lb) 1,376 kg (3,033 lb) 1,183 kg (2,607 lb)
Launch Date	
PMAs 1 and 2	Dec. 4, 1998 STS-88/ISS-2A
PMA 3	Oct. 11, 2000 STS-92/ISS-3A

Nodes

Node 1 (Unity), Node 2, Node 3

NASA/Boeing, Alcatel Alenia Space

Nodes are U.S. modules that connect the elements of the ISS. Node 1, called Unity, was the first U.S.-built element of the ISS that was launched, and it connects the U.S. and Russian segments of the ISS.

Node 2 will connect the U.S., European, and Japanese laboratories. Node 3, still in development, will provide additional habitation functions, including hygiene and sleeping compartments. Nodes 2 and 3 are slightly longer than Node 1.

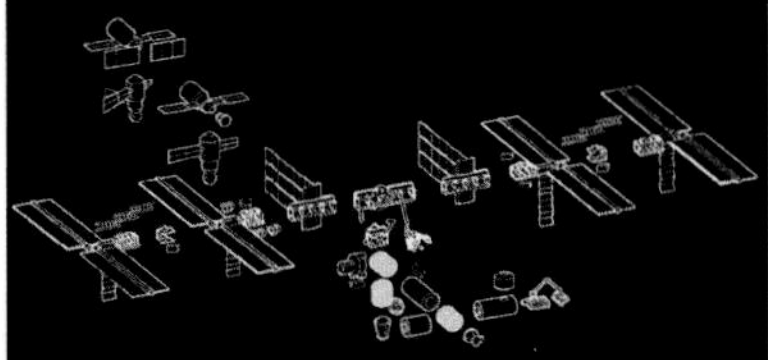

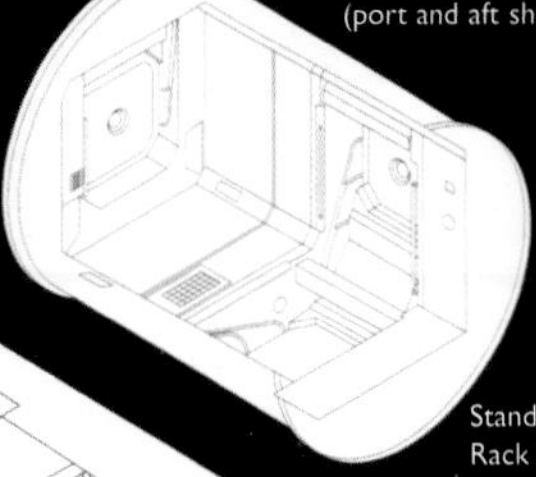

Interior of Node 1 deck (port and aft shown).

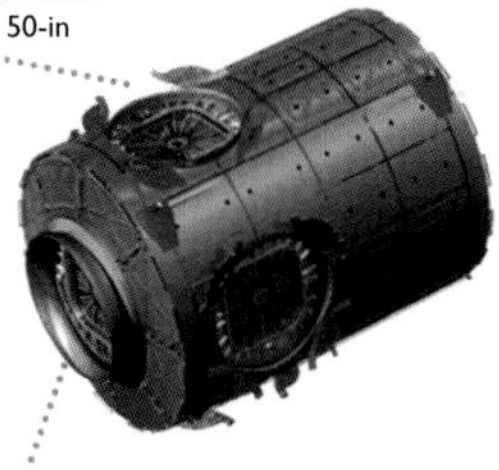

Interior view of Node 1.

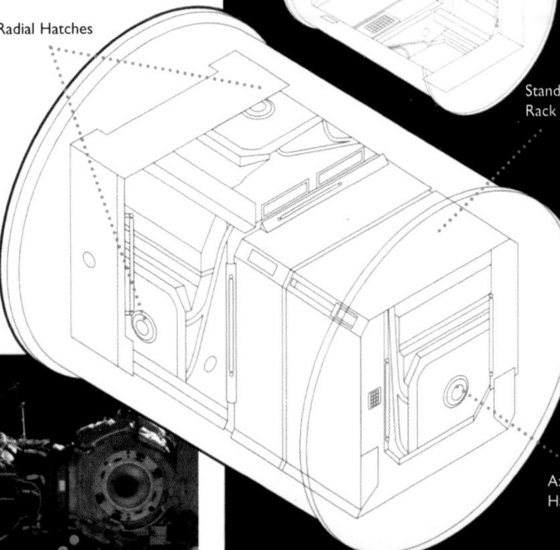

Interior of Node 1 (ceiling, port, and forward endcone shown).

Node 1 is shown with the Russian segment FGB to the right (aft), the U.S. Lab to the left (fore), the U.S. Airlock at the bottom (starboard), and PMA 3 at the top (port).

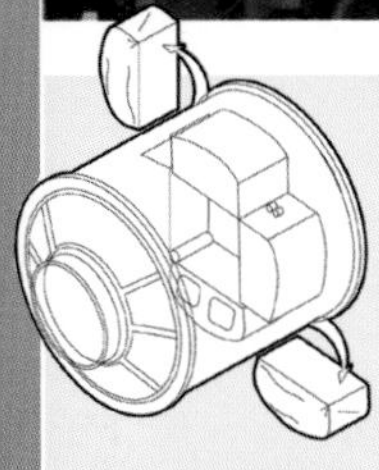

Placement of 4 racks in Node 1.

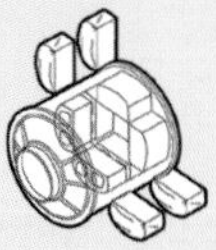

Nodes 2 and 3 have locations for 8 racks.

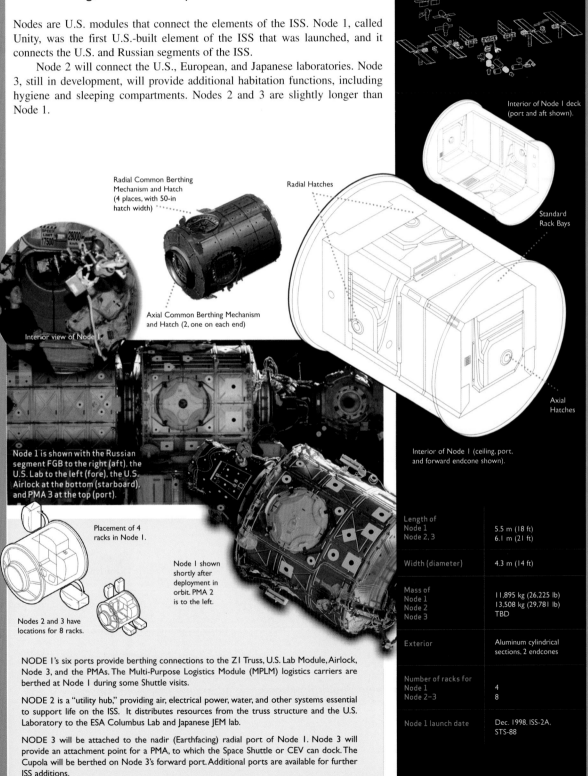

Node 1 shown shortly after deployment in orbit. PMA 2 is to the left.

Length of Node 1 Node 2, 3	 5.5 m (18 ft) 6.1 m (21 ft)
Width (diameter)	4.3 m (14 ft)
Mass of Node 1 Node 2 Node 3	 11,895 kg (26,225 lb) 13,508 kg (29,781 lb) TBD
Exterior	Aluminum cylindrical sections, 2 endcones
Number of racks for Node 1 Node 2–3	 4 8
Node 1 launch date	Dec. 1998, ISS-2A, STS-88

NODE 1's six ports provide berthing connections to the Z1 Truss, U.S. Lab Module, Airlock, Node 3, and the PMAs. The Multi-Purpose Logistics Module (MPLM) logistics carriers are berthed at Node 1 during some Shuttle visits.

NODE 2 is a "utility hub," providing air, electrical power, water, and other systems essential to support life on the ISS. It distributes resources from the truss structure and the U.S. Laboratory to the ESA Columbus Lab and Japanese JEM lab.

NODE 3 will be attached to the nadir (Earthfacing) radial port of Node 1. Node 3 will provide an attachment point for a PMA, to which the Space Shuttle or CEV can dock. The Cupola will be berthed on Node 3's forward port. Additional ports are available for further ISS additions.

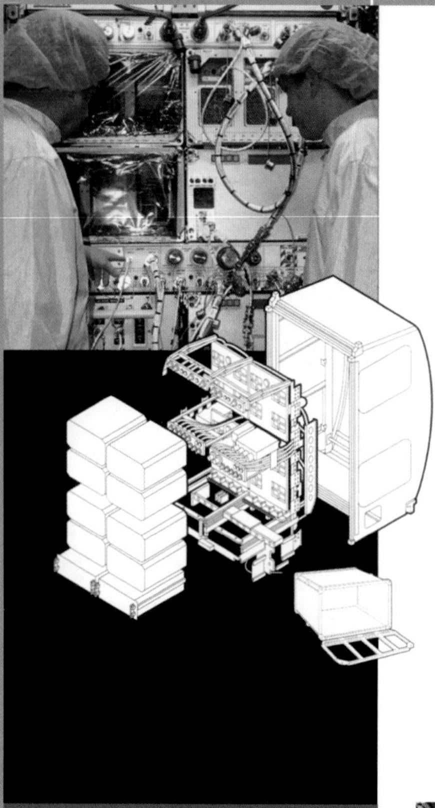

Power	
3, 6, or 12 kW, 114.5–126 voltage, direct current (VDC)	
Data	
Low rate	MIL-STD-1553 bus 1 Mbps
High rate	100 Mbps
Ethernet	10 Mbps
Video	NTSC
Gases	
Nitrogen	Flow = 0.1 kg/min minimum 517–827 kPa, nominal 1,379 kPa, maximum
Argon, carbon dioxide, helium	517–768 kPa, nominal 1,379 kPa, maximum
Cooling Loops	
Moderate temperature	16.1 °C–18.3 °C
Flow rate	0–45.36 kg/h
Low temperature	3.3 °C–5.6 °C
Flow rate	233 kg/h
Vacuum	
Venting	10–3 torr in less than 2 h for single payload of 100 L
Vacuum resource	10–3 torr

Internal Research Accommodations

Several research facilities are in place aboard the Station to support science investigations.

Standard Payload Racks

Research payloads within the U.S., European, and Japanese laboratories typically are housed in a standard rack, such as the International Standard Payload Rack (ISPR). Smaller payloads may fit in a Shuttle middeck locker equivalent and be carried in a rack framework.

Active Rack Isolation System (ARIS)

The ARIS is designed to isolate payload racks from vibration. The ARIS is an active electromechanical damping system attached to a standard rack that senses the vibratory environment with accelerometers and then damps it by introducing a compensating force.

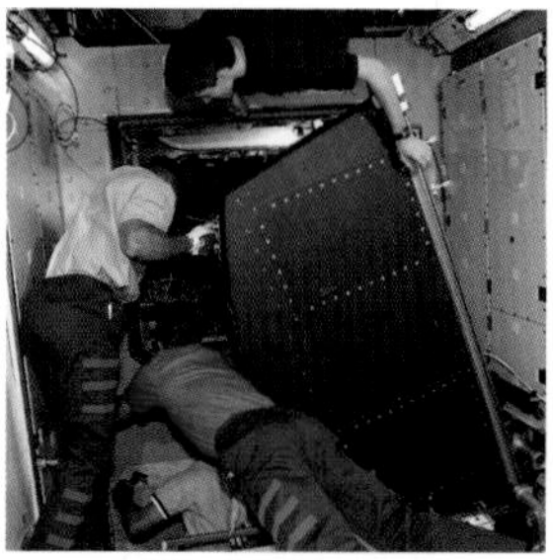

Crew installs a rack in the U.S. Lab in orbit.

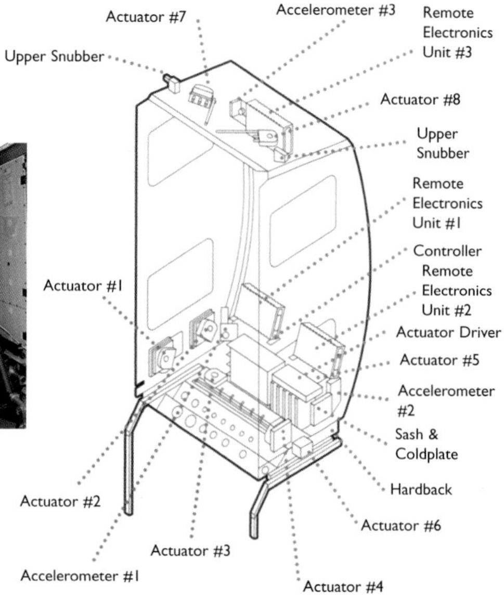

Research Rack Locations

INTERNATIONAL PRESSURIZED SITES	STATION-WIDE	U.S. SHARED
U.S. Laboratory	13	13
Japanese Experiment Module	11	5
European Columbus Research Laboratory	10	5
Total	34	23

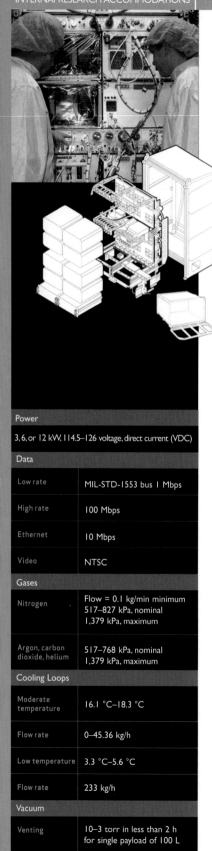

Installation of a rack in the U.S. Lab prior to launch.

External Research Accommodations

Many locations are available for the mounting of payloads or experiments on the outside of the Station: on the U.S. Truss, on the Russian elements, and additional accommodations will be provided when the Japanese Experiment Module (JEM) Exposed Facility (EF) and Columbus modules are attached.

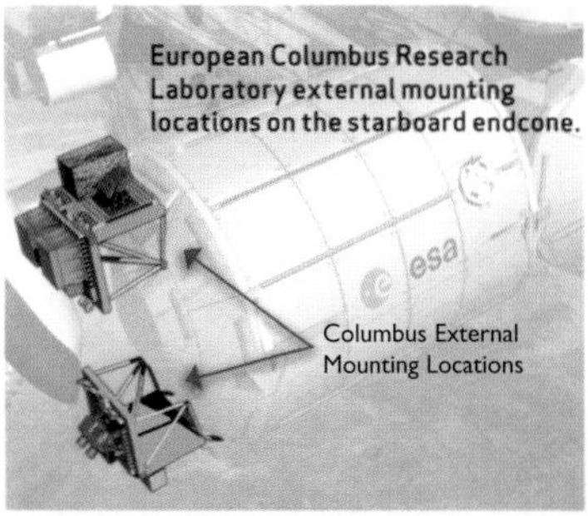

JEM-EF in preparation for launch.

External Research Locations

EXTERNAL UNPRESSURIZED ATTACHMENT SITES	STATION-WIDE	U.S. SHARED
U.S. Truss	10	10
Japanese Exposed Facility	10	5
European Columbus Research Laboratory	4	0
Total	24	15

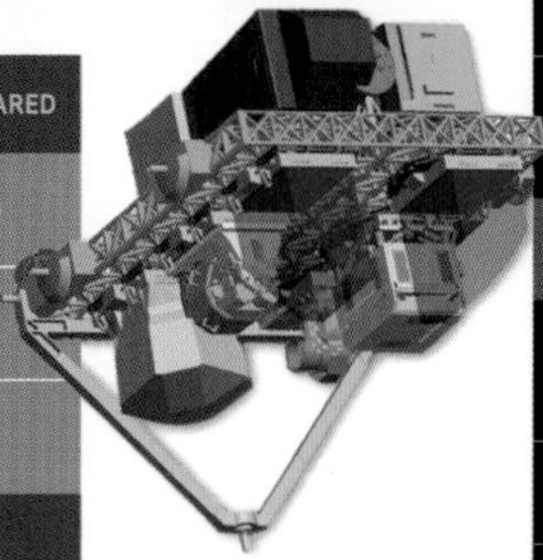

External Payload Accommodations

External payloads may be accommodated at several locations on the U.S. S3 and P3 Truss segments. External payloads are accommodated on an Expedite the Processing of Experiments to the Space Station racks (EXPRESS) Logistics Carrier (ELC). Mounting spaces are provided, and interfaces for power and data are standardized to provide quick and straightforward payload integration. Payloads can be mounted using the Special Purpose Dexterous Manipulator (SPDM), Dextre, on the Station's robotic arm.

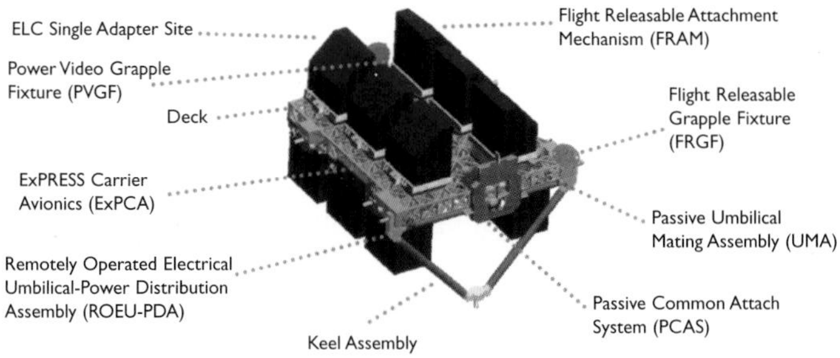

Express Logistics Carrier (ELC) Resources	
Mass capacity	4,445 kg (9,800 lb)
Volume	30 m^3
Power	3 kW maximum, 113-126 VDC
Low-rate data	1 Mbps (MIL-STD-1553)
High-rate data	95 Mbps (shared)
Local area network	6 Mbps (802.3 Ethernet)

ELC Single Adapter Resources	
Mass capacity	227 kg (500 lb)
Volume	1 m^3
Power	750 W, 113-126 VDC 500 W at 28 VDC per adapter
Thermal	Active heating, passive cooling
Low-rate data	1 Mbps (MIL-STD-1553)
Medium-rate data	6 Mbps (shared)

JEM-EF Resources	
Mass capacity	550 kg (1,150 lb) at standard site 2,250 kg (5,550 lb) at large site
Volume	1.5 m^3
Power	3-6 kW, 113-126 VDC
Thermal	3-6 kW cooling
Low-rate data	1 Mbps (MIL-STD-1553)
High-rate data	43 Mbps (shared)

European Columbus Research Laboratory Resources	
Mass capacity	230 kg (500 lb)
Volume	1 m^3
Power	2.5 kW total to carrier (shared)
Thermal	Passive
Low-rate data	1 Mbps (MIL-STD-1553)
Medium-rate data	2 Mbps (shared)

U.S. Laboratory Module (Destiny)

NASA/Boeing

The U.S. Lab provides internal interfaces to accommodate the resource requirements of 24 equipment racks. Approximately half of these are for accommodation and control of ISS systems, and the remainder support scientific research.

Destiny was the first research module installed on the Station. The side of Destiny that usually faces Earth contains a large circular window of very high optical quality.

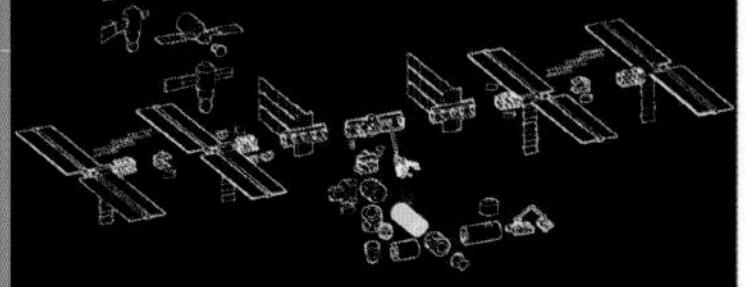

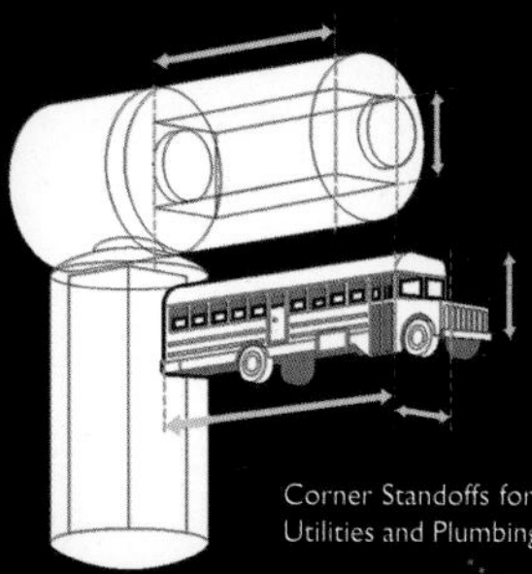

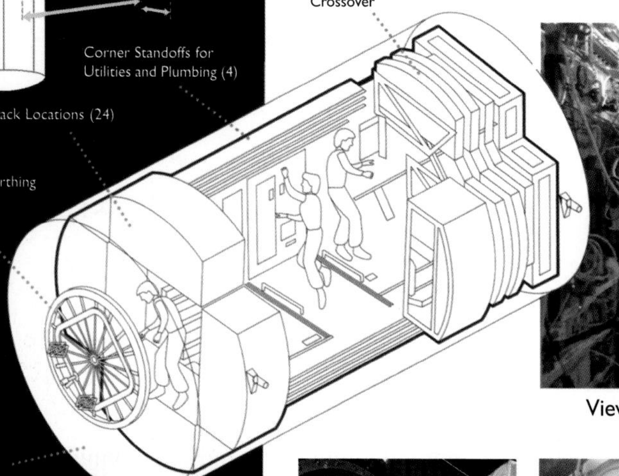

View of astronaut Ed Lu in the U.S. Lab.

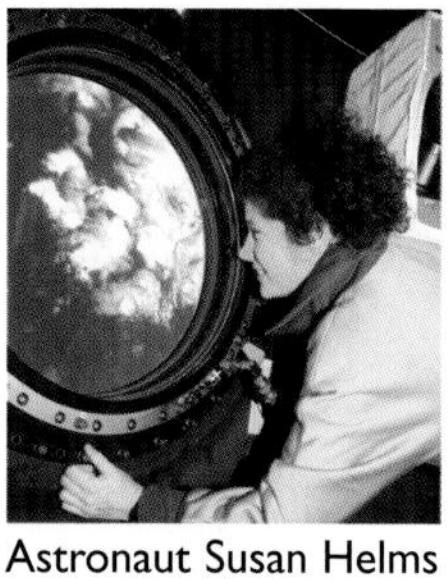

Astronaut Susan Helms at the 20-inch-diameter circular window.

Module in preparation at Kennedy Space Center (KSC).

Length	8.5 m (28 ft)
Length with attached Common Berthing Mechanism (CBM)	9.2 m (30.2 ft)
Width	4.3 m diameter (14 ft)
Mass	14,515 kg (32,000 lb) 24,023 kg (52,962 lb) with all racks and outfitting
Exterior	Aluminum, 3 cylindrical sections, 2 endcones
Number of racks	24 (13 scientific and 11 system)
Windows	1, with a diameter of 50.9 cm (20 in)
Launch date	Feb. 7, 2001, assembly flight 5A, STS-98

U.S. Lab after deployment. The Pressurized Mating Adapter (PMA) is

The Human Research Facility (HRF) supports a variety of life sciences experiments. It includes equipment for lung function tests, ultrasound equipment to image the heart, and many other types of computers and medical equipment.

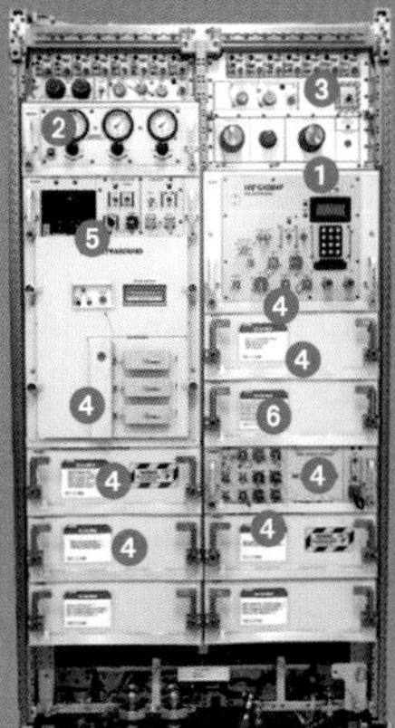

1 Gas Analyzer System for Metabolic Analysis Physiology (GASMAP)
2 GASMAP Gas Calibration Module (GCM)
3 Power Switch and Data Interconnects
4 Stowage Drawers
5 Ultrasound Imaging System
6 Workstation Interface

John Phillips conducts Foot Reaction Forces (FOOT) experiment on HRF rack.

The Microgravity Science Glovebox provides a sealed environment for conducting science and technology experiments. It has a large front window and built-in gloves, data storage and recording capabilities, and an independent air circulation and filtration system.

1 Airlock
2 Control and Monitoring Panel
3 Power Distribution Box
4 Power Switches
5 Remote Power Distribution
6 Work Volume Armholes
7 Video

William McArthur uses the Microgravity Science Glovebox.

The five ExPRESS Racks provide subrack-sized experiments with standard utilities such as power, data, cooling, fluids, and gases. The racks stay in orbit, while experiments are changed as needed.

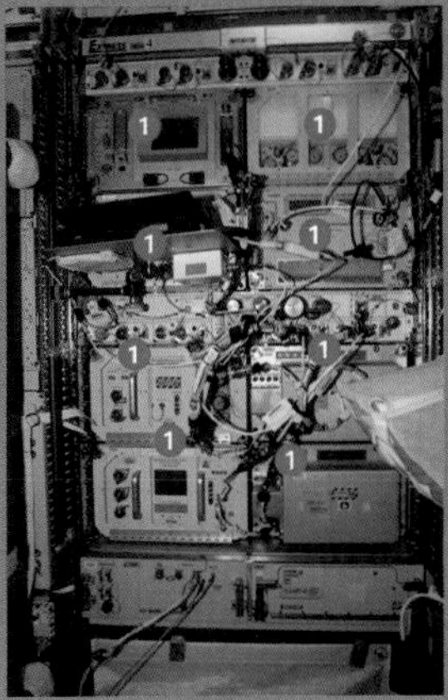

1 Stowage or Payload Locations

The Minus Eighty-Degree Laboratory Freezer for ISS (MELFI) provides refrigerated storage and fast-freezing of biological and life science samples. It can hold up to 300 L of samples ranging in temperature from 4 °C to a low of -80 °C.

1 Refrigerated/Frozen Storage Dewars

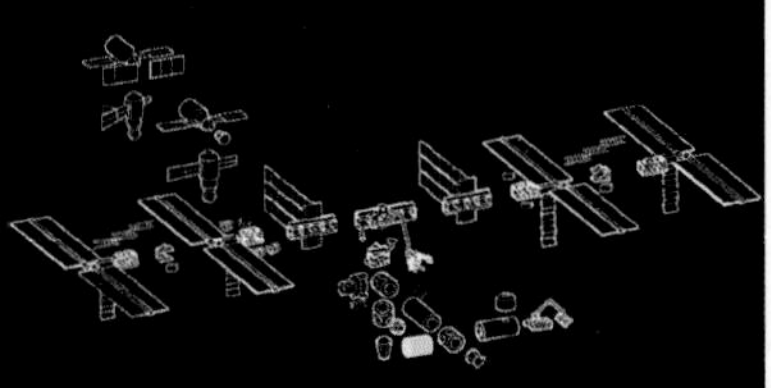

Columbus Research Laboratory

European Space Agency (ESA)/European Aeronautic Defence and Space Co. (EADS) Space Transportation

The Columbus Research Laboratory is Europe's largest contribution to the construction of the International Space Station. It will support scientific and technological research in a microgravity environment. Columbus, a program of ESA, is a multifunctional pressurized laboratory that will be permanently attached to Node 2 of the ISS to carry out experiments in materials science, fluid physics, and biosciences, as well as to perform a number of technological applications.

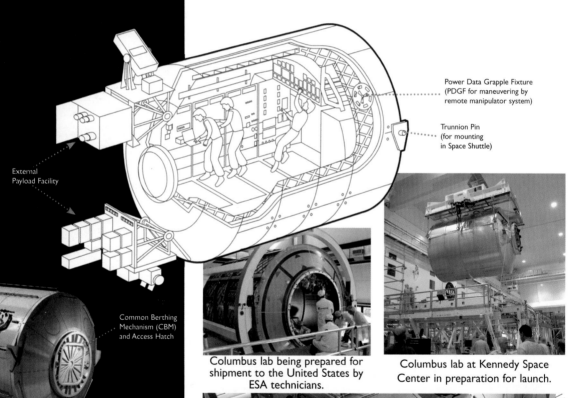

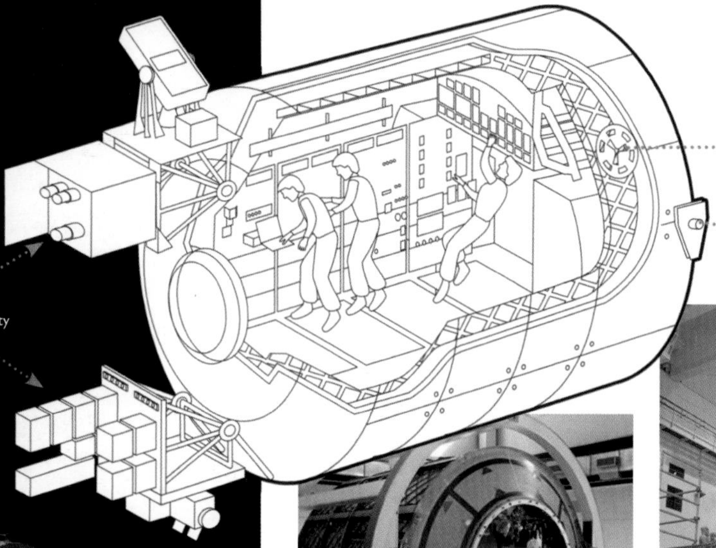

Columbus lab being prepared for shipment to the United States by ESA technicians.

Columbus lab at Kennedy Space Center in preparation for launch.

Columbus berthed to Node 2. PMA 2 at right.

Length	6.9 m (22.6 ft)
Diameter	4.5 m (14.7 ft)
Mass without payload with payload	 10,300 kg (22,700 lb) 19,300 kg (42,550 lb)
Racks	10 International Standard Payload Racks (ISPRs)

Japanese Experiment Module (JEM)/Kibo (Hope)

Japan Aerospace Exploration Agency (JAXA)/Mitsubishi Heavy Industries, Ltd.

The Japanese Experiment Module is the first crewed space facility ever developed by Japan. The Pressurized Module (PM) is used mainly for microgravity experiments. The Exposed Facility (EF) is located outside the pressurized environment of the ISS. Numerous experiments that require direct exposure can be mounted with the help of the JEM remote manipulator and airlock. Logistics components will be launched in the Experiment Logistics Module Pressurized Section (ELM-PS). Experiments may be mounted on the JEM-EF using the Experiment Logistics Module Exposed Section (ELM-ES). All of the JEM modules will be launched on the Space Shuttle.

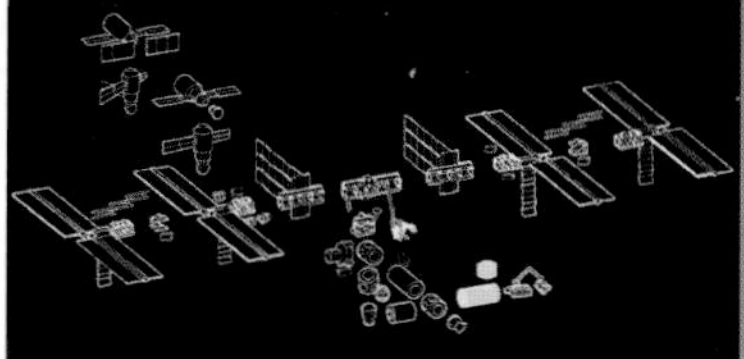

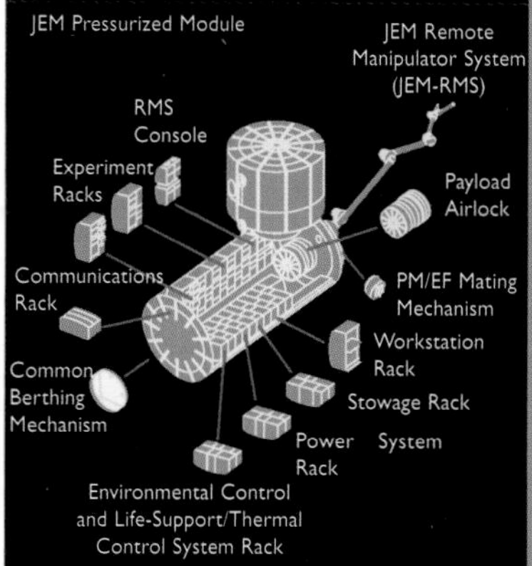

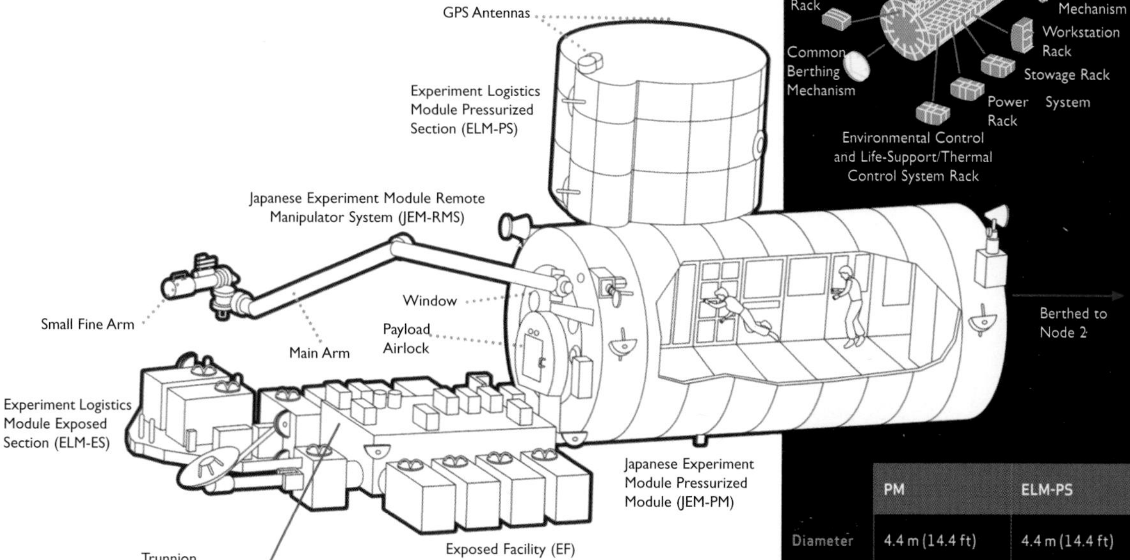

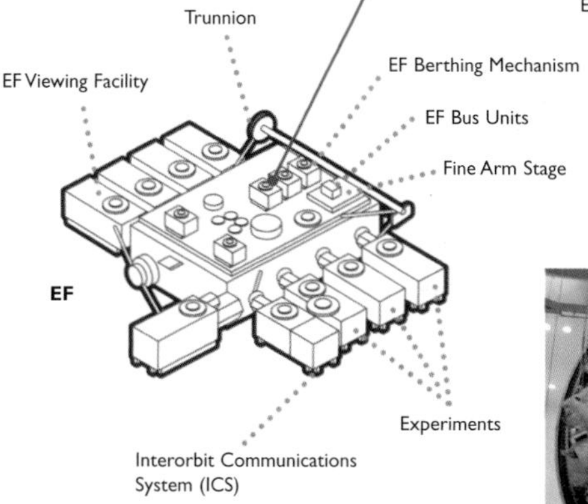

JEM-PM during testing.

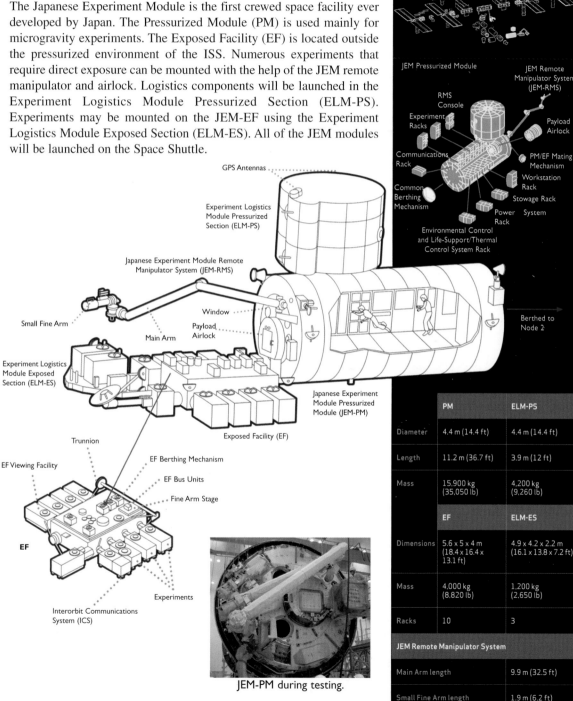

	PM	ELM-PS
Diameter	4.4 m (14.4 ft)	4.4 m (14.4 ft)
Length	11.2 m (36.7 ft)	3.9 m (12 ft)
Mass	15,900 kg (35,050 lb)	4,200 kg (9,260 lb)
	EF	**ELM-ES**
Dimensions	5.6 x 5 x 4 m (18.4 x 16.4 x 13.1 ft)	4.9 x 4.2 x 2.2 m (16.1 x 13.8 x 7.2 ft)
Mass	4,000 kg (8,820 lb)	1,200 kg (2,650 lb)
Racks	10	3
JEM Remote Manipulator System		
Main Arm length		9.9 m (32.5 ft)
Small Fine Arm length		1.9 m (6.2 ft)

Cupola

NASA/Boeing, ESA/Alcatel Alenia Space

The Cupola (named after the raised observation deck on a railroad caboose) is a small module designed for the observation of operations outside the ISS such as robotic activities, the approach of vehicles, and extravehicular activity (EVA). It will also provide spectacular views of Earth and celestial objects. The Cupola has six side windows and a top window, all of which are equipped with shutters to protect them from contamination and collisions with orbital debris or micrometeorites. The Cupola is designed to house computer workstations that control the ISS and the remote manipulators. It can accommodate two crewmembers simultaneously and is berthed to a Node using the Common Berthing Mechanism (CBM).

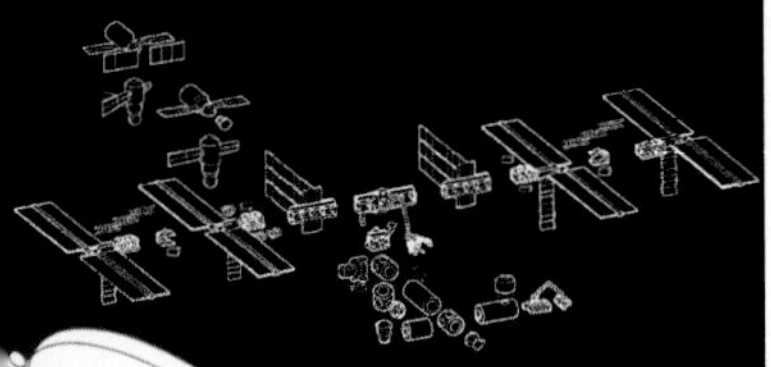

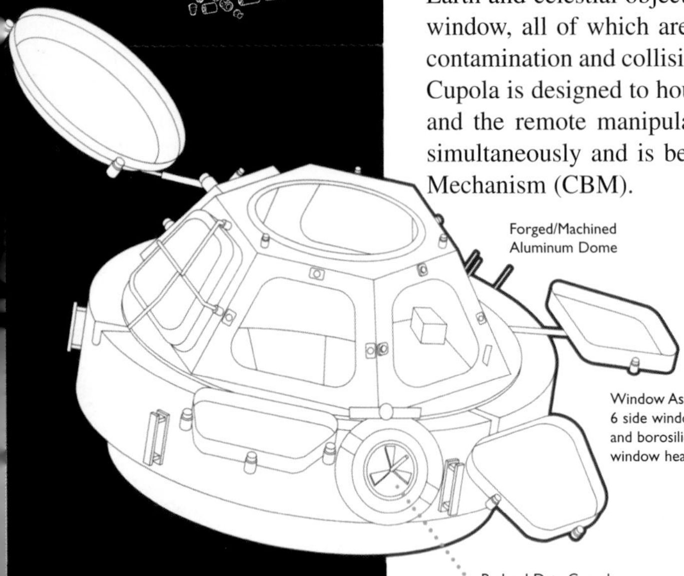

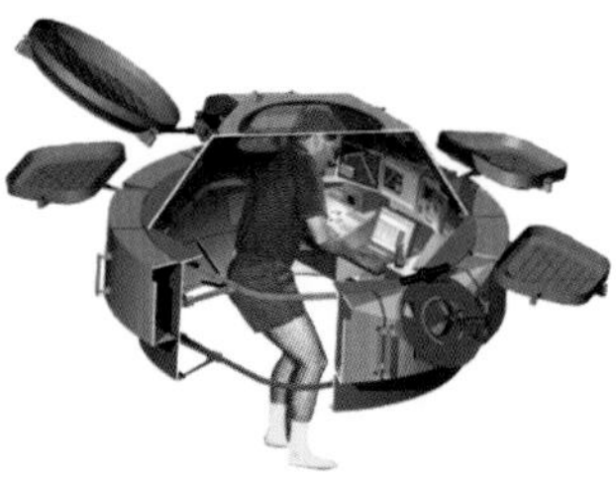

Length	3 m (9.8 ft)
Height	1.5 m (4.7 ft)
Diameter	3 m (9.8 ft)
Mass	1,880 kg (4,136 lb)
Capacity	2 crewmembers with portable workstation

Command and control workstation based on portable computer system.

The Cupola in development.

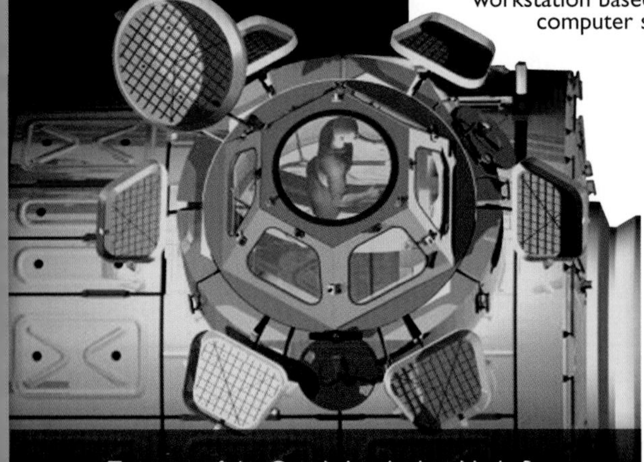

Top view of the Cupola berthed to Node 3.

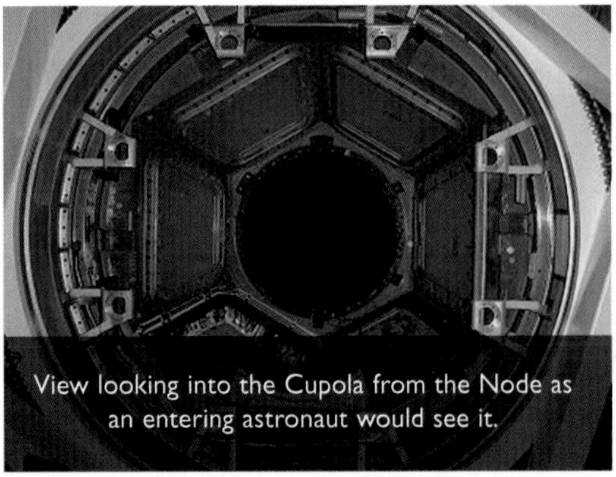

View looking into the Cupola from the Node as an entering astronaut would see it.

Mobile Servicing System (MSS)

Space Station Remote Manipulator System (SSRMS) and Special Purpose Dexterous Manipulator (SPDM/Dextre)

Mobile Base System (MBS), Canadian Space Agency (CSA)/ MacDonald, Dettwiler and Associates, Ltd.

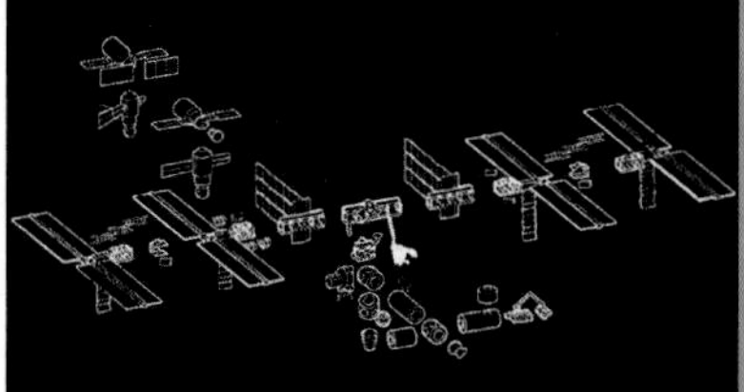

The Mobile Servicing System (MSS) plays a key role in the construction of the ISS and general Station operations. It allows astronauts and cosmonauts to work from inside the Station, thus reducing the number of spacewalks. The MSS Operations Complex in Longueuil, Quebec, is the ground base for the system.

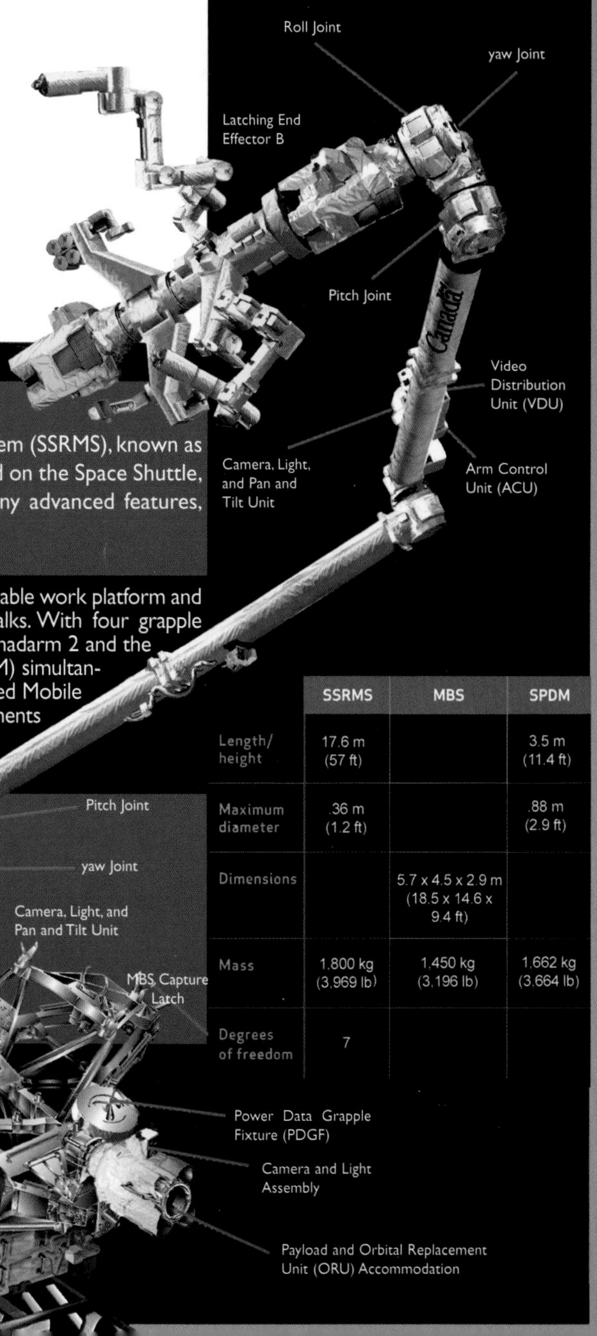

The MSS has three parts:

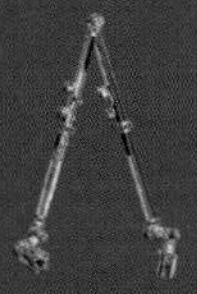

The Space Station Remote Manipulator System (SSRMS), known as Canadarm 2, is similar to the Canadarm used on the Space Shuttle, but Canadarm 2 is larger, incorporates many advanced features, and includes the ability to self-relocate.

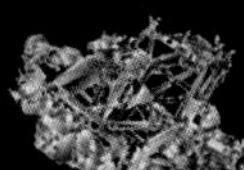

The Mobile Base System (MBS) provides a movable work platform and storage facility for astronauts during spacewalks. With four grapple fixtures, it can serve as a base for both the Canadarm 2 and the Special Purpose Dexterous Manipulator (SPDM) simultaneously. Since it is mounted on the U.S.-provided Mobile Transporter (MT), the MBS can move key elements to their required worksites by moving along a track system mounted on the ISS truss.

The Special Purpose Dexterous Manipulator (SPDM) has a dual arm design that can remove and replace smaller components on the Station's exterior, where precise handling is required. It will be equipped with lights, video equipment, and a tool platform, as well as four tool holders.

	SSRMS	MBS	SPDM
Length/ height	17.6 m (57 ft)		3.5 m (11.4 ft)
Maximum diameter	.36 m (1.2 ft)		.88 m (2.9 ft)
Dimensions		5.7 x 4.5 x 2.9 m (18.5 x 14.6 x 9.4 ft)	
Mass	1,800 kg (3,969 lb)	1,450 kg (3,196 lb)	1,662 kg (3,664 lb)
Degrees of freedom	7		

SSRMS during testing.

Scott Parazynski removes orbital debris shield to connect SSRMS wiring.

transportation/ logistics

Building and maintaining the International Space Station (ISS) is a very complex task. An international fleet of space vehicles launches ISS components; rotates crews; provides logistical support; and replenishes propellant, items for science experiments, and other necessary supplies and equipment. The Space Shuttle must be used to deliver most ISS modules and major components.

All of these important deliveries sustain a constant supply line that is crucial to the development and maintenance of the International Space Station. The fleet is also responsible for returning experiment results to Earth and for removing trash and waste from the ISS.

Currently, transport vehicles are launched from two sites on Earth. In the future, the number of launch sites will increase to four or more. Future plans also include new commercial transports that will take over the role of U.S. ISS logistical support.

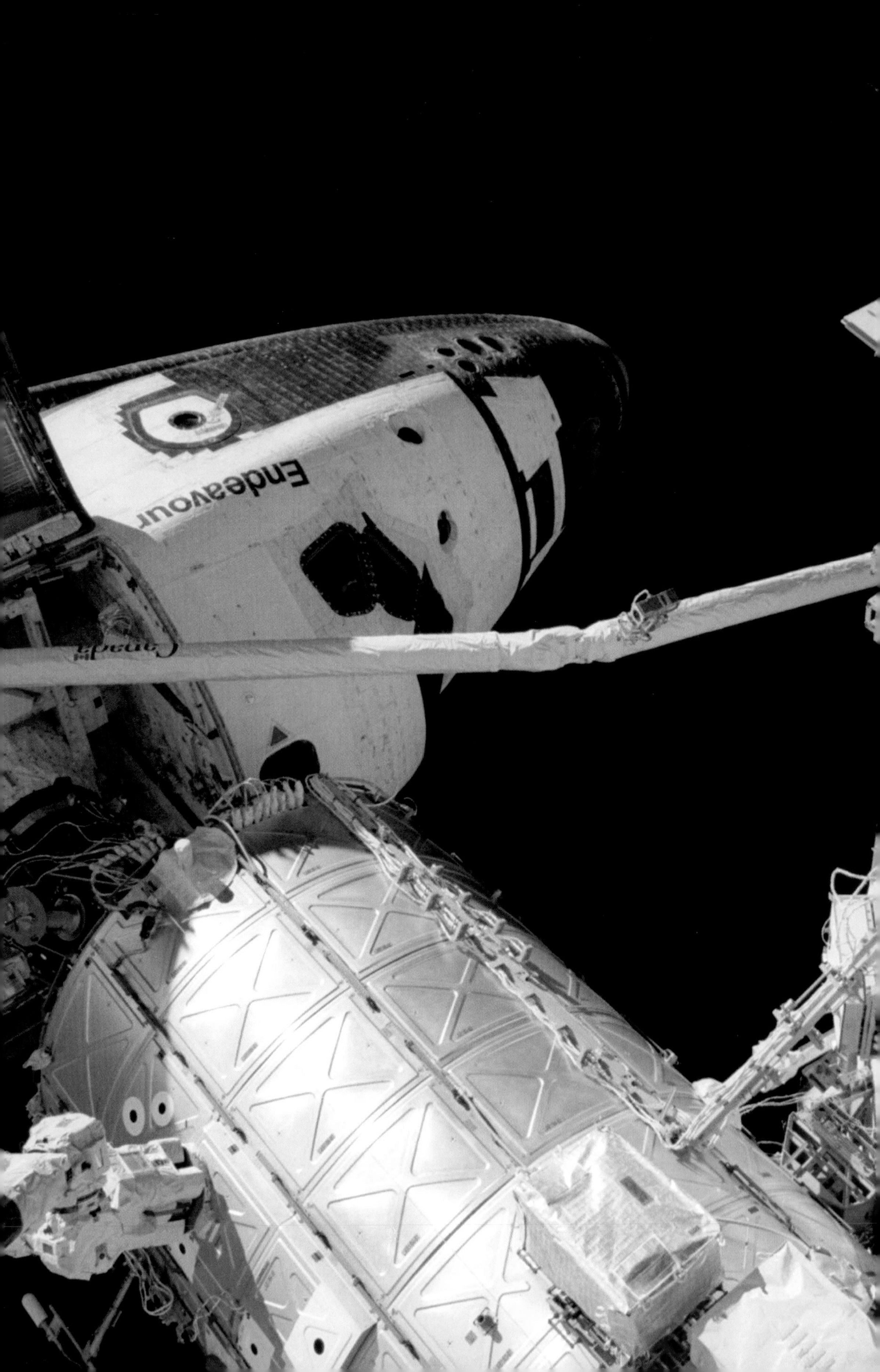
Endeavour

Soyuz Proton

Roscosmos
Russia

H-II

JAXA
Japan

Ariane

ESA
Europe

Shuttle

NASA
United States

	RUSSIA		JAPAN	EUROPE	U.S.
	Soyuz SL-4	Proton SL-12	H-II	Ariane 5	Space Shuttle
First launch	1957 1963 (Soyuz variant)	1965	1996	1996	1981
Launch site(s)	Baikonur Cosmodrome	Baikonur Cosmodrome	Tanegashima Space Center	Guiana Space Center	Kennedy Space Center
Launch performance payload capacity	7,150 kg (15,750 lb)	20,000 kg (44,000 lb)	16,500 kg (36,400 lb)	18,000 kg (39,700 lb)	18,600 kg (41,000 lb) 105,000 kg (230,000 lb), orbiter only
Return performance payload capacity	N/A	N/A	N/A	N/A	18,600 kg (41,000 lb) 105,000 kg (230,000 lb), orbiter only
Number of stages	2 + 4 strap-ons	4 + 6 strap-ons	2 + 2 strap-ons	2 + 2 strap-ons	1.5 + 2 strap-ons
Length	49.5 m (162 ft)	57 m (187 ft)	53 m (173 ft)	51 m (167 ft)	56.14 m (18.2 ft) 37.24 m (122.17 ft), orbiter only
Mass	310,000 kg (683,400 lb)	690,000 kg (1,521,200 lb)	570,000 kg (1,256,600 lb)	746,000 kg (1,644,600 lb)	2,040,000 kg (4,497,400 lb)
Launch thrust	6,000 kN (1,348,800 lbf)	9,000 kN (2,023,200 lbf)	5,600 kN (1,258,900 lbf)	11,400 kN (2,562,820 lbf)	34,677 kN (7,795,700 lbf)
Payload Examples	Soyuz Progress Pirs	Service Module Functional Cargo Block (FGB) Research Module (RM) Multipurpose Lab Module (MLM)	H-II Transfer Vehicle (HTV)	Ariane Automated Transfer Vehicle (ATV)	Shuttle Orbiter Nodes, U.S. Lab Columbus, JEM, Truss elements Airlock, SSRMS

The largest U.S. and Russian launch vehicles are used to place elements of the ISS, crew, and cargo in orbit. Eventually, Japanese and European launch vehicles will support cargo delivery. Currently, only the U.S. Space Shuttle provides the capability to return significant payloads.

Soyuz

S.P. Korolev Rocket and Space Corporation Energia (RSC Energia)

Soyuz spacecraft have been in use since the mid-1960s and have been upgraded periodically. Soyuz can support three suited crewmembers for up to 3 days. A nitrogen/oxygen atmosphere at sea level pressure is provided. The vehicle has an automatic docking system and may be piloted automatically or by a crewmember. The Soyuz TMA used for the ISS includes changes to accommodate larger and smaller crewmembers, an improved landing system, and digital electronic controls and displays.

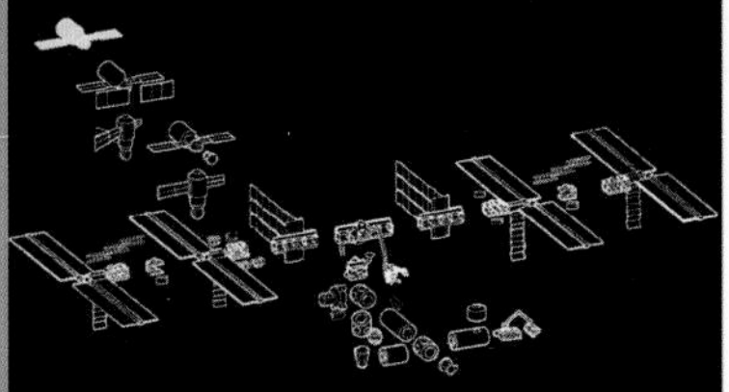

Soyuz departs ISS.

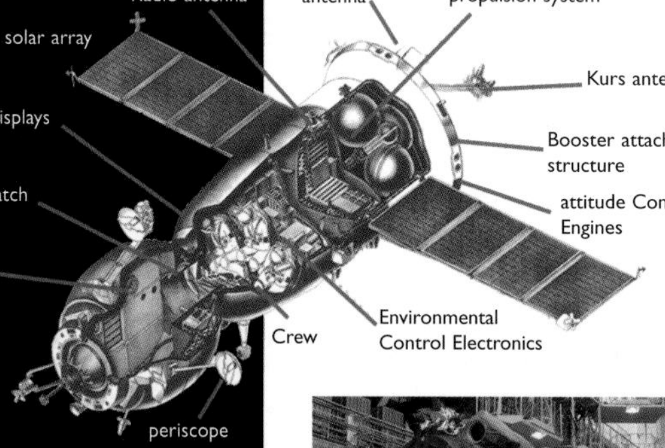

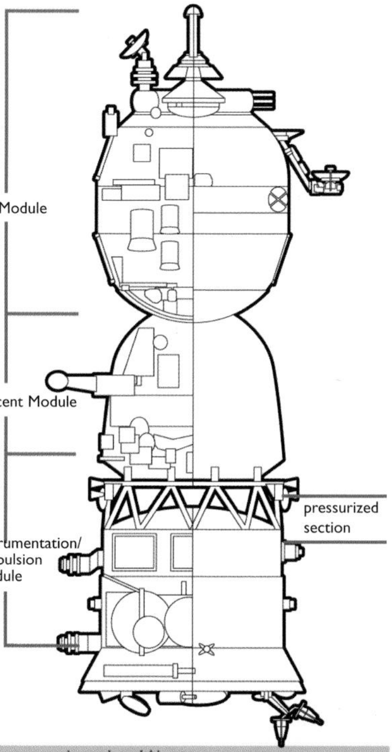

Soyuz being prepared for launch.

Soyuz descent module interior.

Launch mass	6,441 kg (14,200 lb)
Descent module	2,630 kg (5,800 lb)
Orbital module	1,179 kg (2,600 lb)
Instrumentation/ propulsion module	2,360 kg (5,200 lb)
Delivered payload (with three crewmembers)	30 kg (66 lb)
Returned payload	50 kg (110 lb)
Length	7 m (22.9 ft)
Maximum diameter	2.7 m (8.9 ft)
Diameter of habitable modules	2.2 m (7.2 ft)
Solar array span	10.7 m (35.1 ft)
Volume of orbital module	6.5 m³ (229.5 ft³)
Volume of descent module	4 m³ (141.3 ft³)
Descent G-loads	3–4 g
Final landing speed	2 m/s (6.6 ft/s)

Mission Sequence

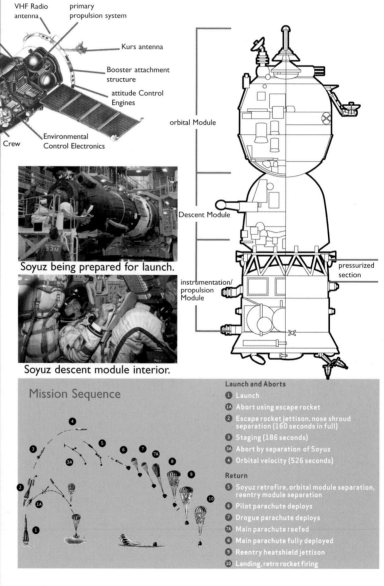

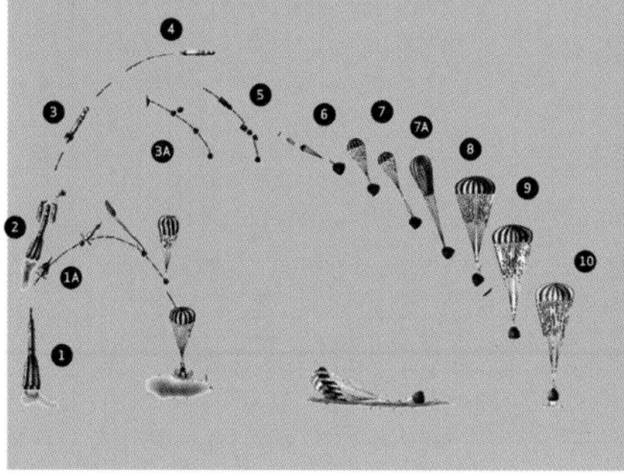

Launch and Aborts

- 1 Launch
- 1A Abort using escape rocket
- 2 Escape rocket jettison, nose shroud separation (160 seconds in full)
- 3 Staging (186 seconds)
- 3A Abort by separation of Soyuz
- 4 Orbital velocity (526 seconds)

Return

- 5 Soyuz retrofire, orbital module separation, reentry module separation
- 6 Pilot parachute deploys
- 7 Drogue parachute deploys
- 7A Main parachute reefed
- 8 Main parachute fully deployed
- 9 Reentry heatshield jettison
- 10 Landing, retro rocket firing

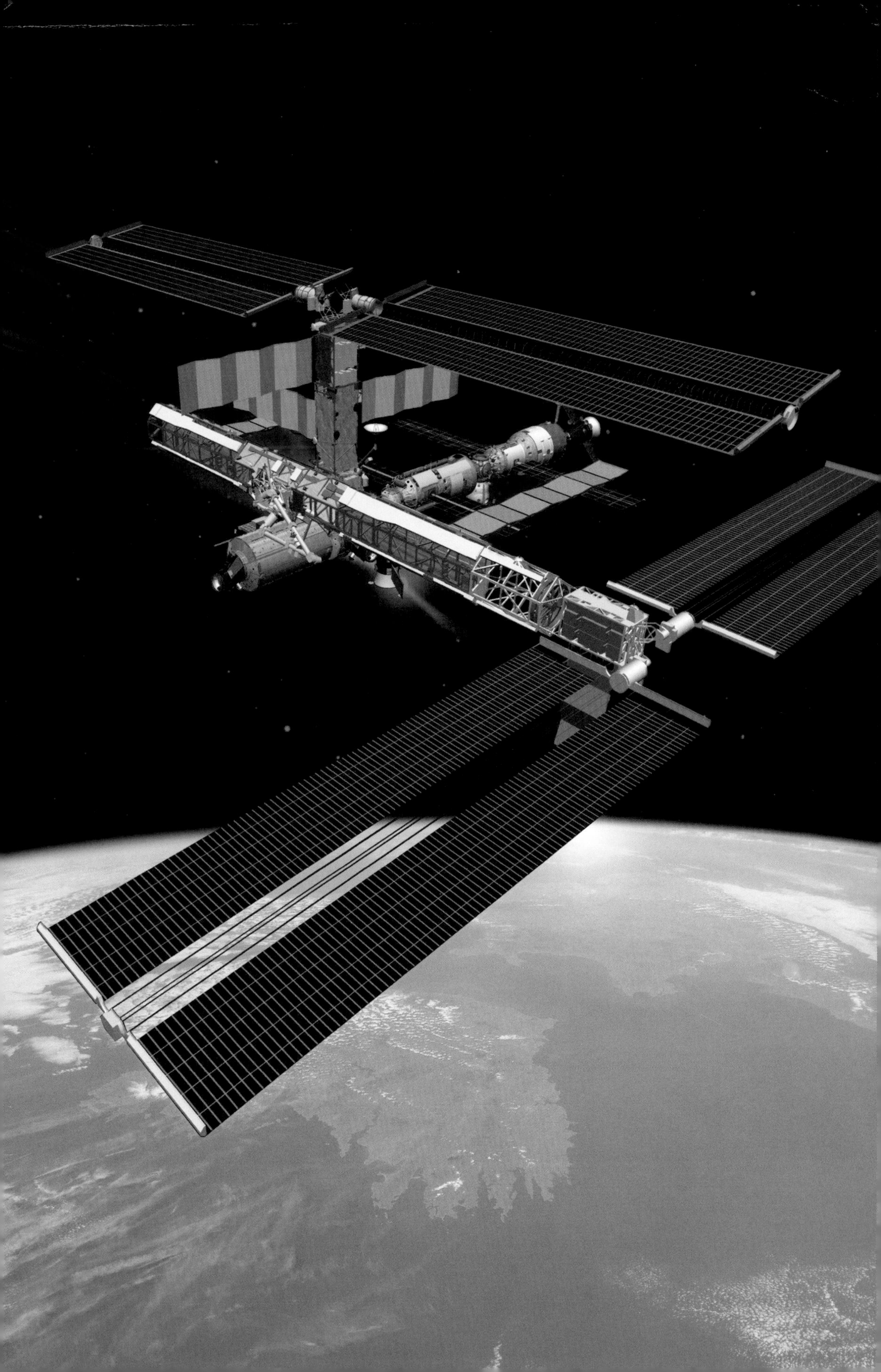

interesting facts

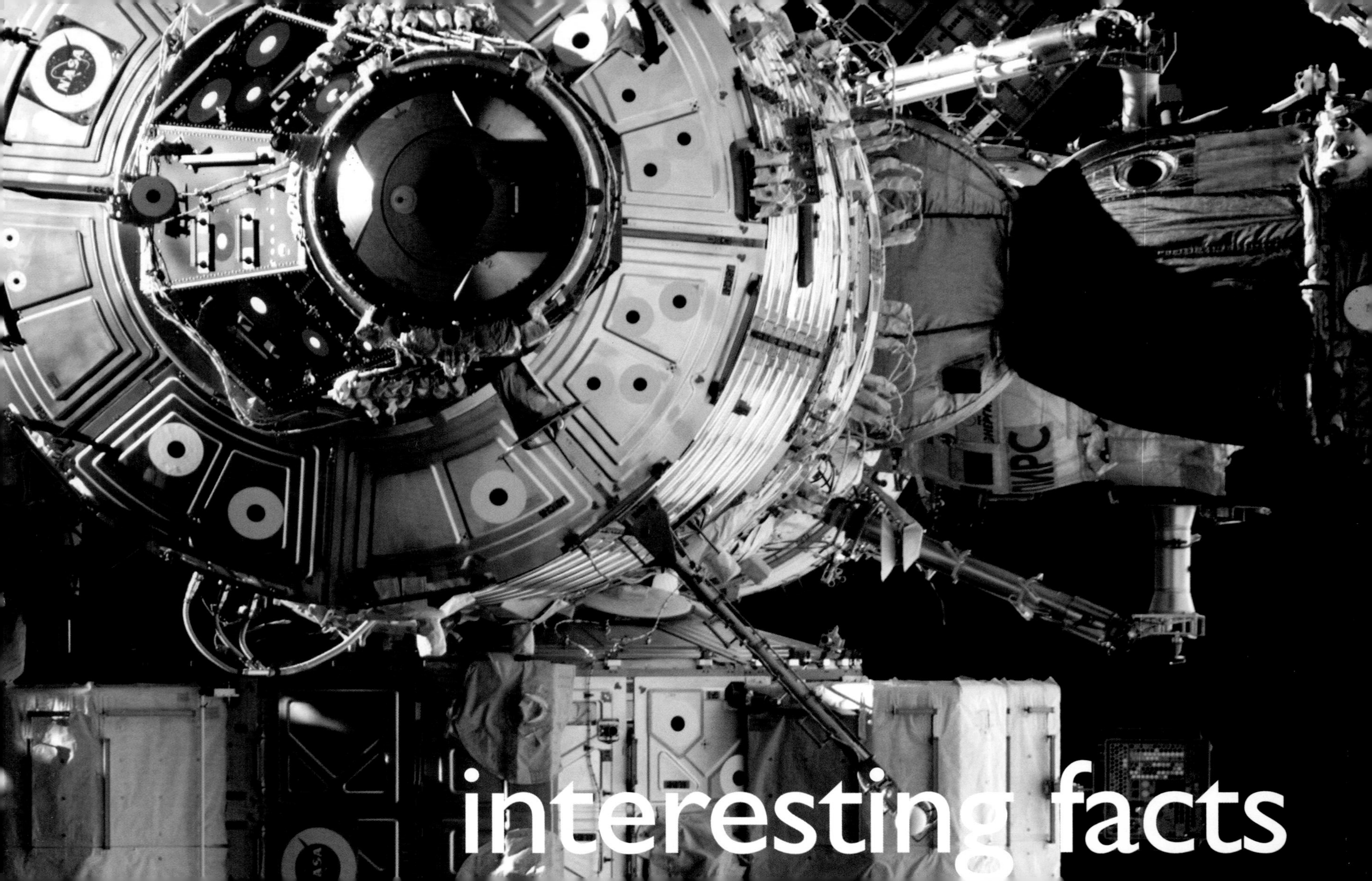

Progress

S.P. Korolev Rocket and Space Corporation Energia (RSC Energia)

Progress is a resupply vehicle used for cargo and propellant deliveries to the ISS. Once docked to the ISS, Progress engines can boost the ISS to higher altitudes and control the orientation of the ISS in space. Typically, three Progress vehicles bring supplies to the ISS each year. Progress is based upon the Soyuz design, and it can either work autonomously or can be flown remotely by crewmembers aboard the ISS. After a Progress vehicle is filled with trash from the ISS, and after undocking and deorbit, it is incinerated in Earth's atmosphere at the end of its mission.

Progress approaches ISS.

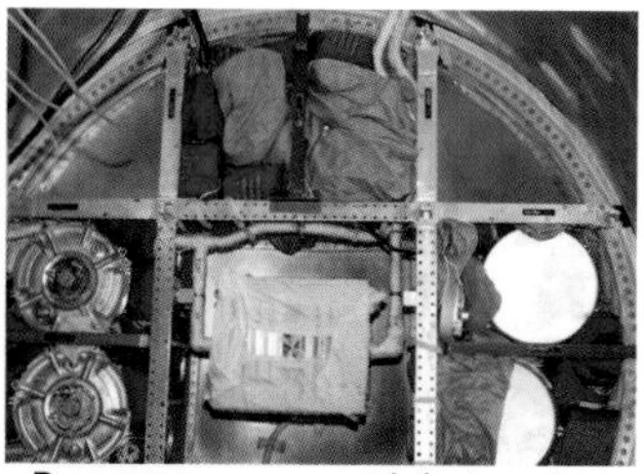

Progress cargo module interior.

Progress prelaunch processing.

Progress prior to reentry.

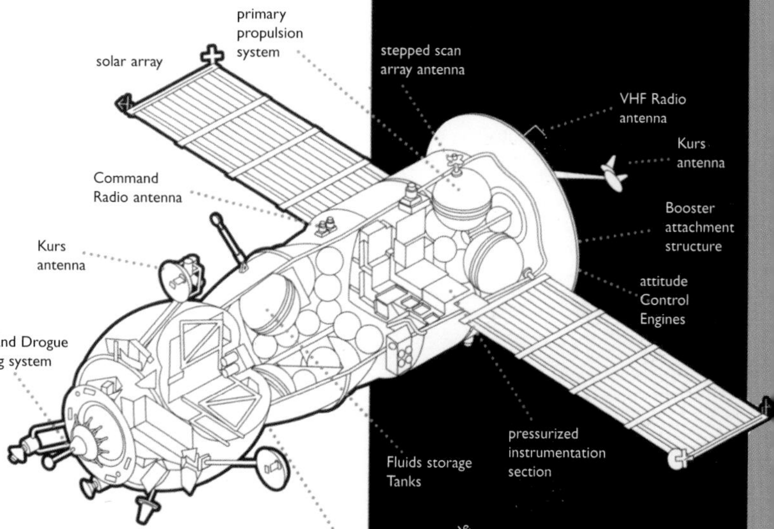

Length	7.4 m (24.3 ft)
Maximum diameter	2.7 m (8.9 ft)
Span with solar arrays	10.6 m (34.8 ft)
Launch mass	7,150 kg (15,800 lb)
Cargo upload capacity	2,230–3,200 kg (4,915–7,055 lb)
Pressurized habitable volume	6.6 m^3 (233 ft^3)
Engine thrust	2,942 N (661 lbf)
Orbital life	6 mo

Cargo Load

	MAXIMUM	TYPICAL*
Dry cargo such as bags	1,800 kg (3,968 lb)	1,070 kg (2,360 lb)
Water	420 kg (925 lb)	300 kg (660 lb)
Air	50 kg (110 lb)	47 kg (103 lb)
Refueling propellant	1,700 kg (3,748 lb)	870 kg (1,918 lb)
Reboost propellant	250 kg (550 lb)	250 kg (550 lb)
Waste capacity	2,000 kg (4,409 lb)	2,000 kg (4,409 lb)

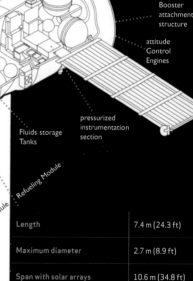

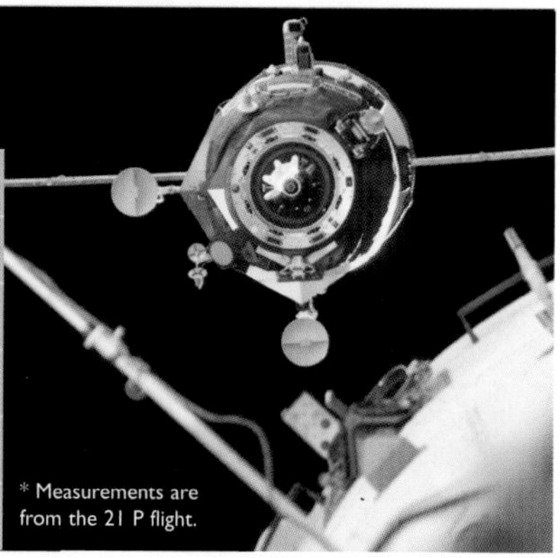

* Measurements are from the 21 P flight.

Space Shuttle Orbiter/Discovery, Atlantis, Endeavour

NASA/Boeing/Rockwell

The U.S. Space Shuttle provides Earth-to-orbit and return capabilities and in-orbit support. The diversity of its missions and customers is testimony to the adaptability of its design. As of mid-2006, the Shuttle had flown 115 times. The Shuttle's primary purpose during the remaining 4 years of operation will be to complete the assembly of the ISS. By 2010, it will be retired.

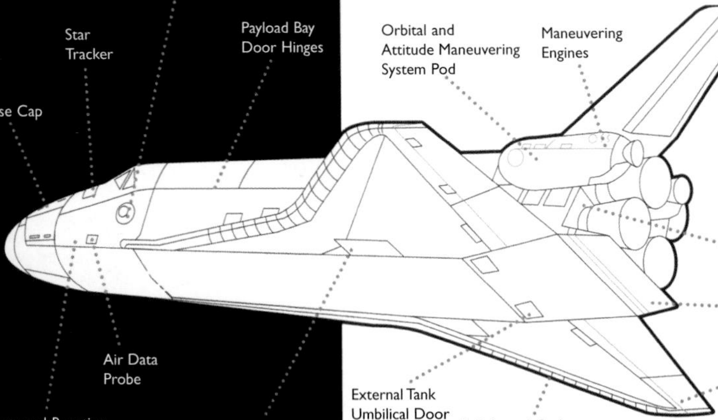

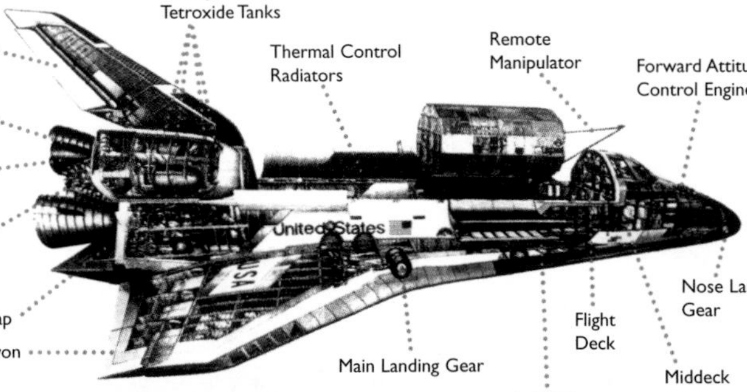

Length	37.2 m (122.2 ft)
Height	17.3 m (56.7 ft)
Wingspan	23.8 m (78 ft)
Typical mass	104,000 kg (230,000 lb)
Cargo capacity	16,000 kg (35,000 lb) (typical launch and return to ISS)
Pressurized habitable volume	74 m³ (2,625 ft³)
Mission length	7–16 days, typical
Number of crew	7, typical
Atmosphere	oxygen-nitrogen
Cargo Bay	
Length	18.3 m (60 ft)
Diameter	4.6 m (15 ft)

The Shuttle approaches the ISS carrying the Multi-Purpose Logistics Module (MPLM).

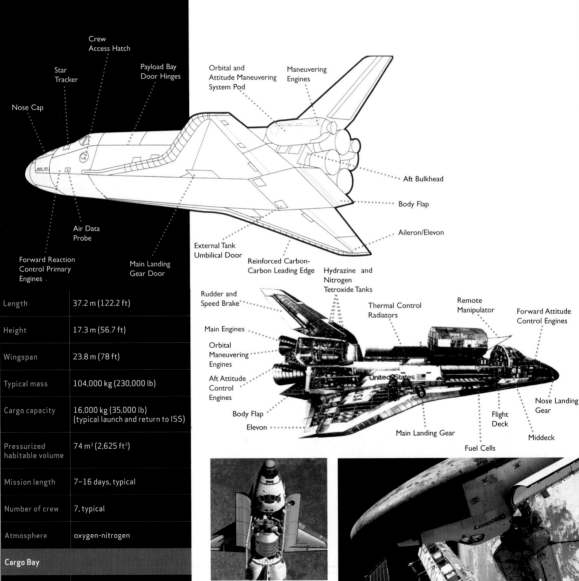

Shuttle berthed at the U.S. Lab, PMA 2.

Multi-Purpose Logistics Module (MPLM)/Leonardo, Raffaelo, Donatello

NASA/Alcatel Alenia Space

The Italian-built Multi-Purpose Logistics Module (MPLM) serves as the International Space Station's "moving van" by carrying laboratory racks filled with equipment, experiments, and supplies to and from the Station aboard the Space Shuttle.

Mounted in the Shuttle's cargo bay for launch and landing, the modules are transferred to the Station using the Shuttle's robotic arm after the Shuttle has docked. While berthed to the Station, racks of equipment and stowage items are unloaded from the module, and racks and equipment may be reloaded to be transported back to Earth. The MPLM is then detached from the Station and positioned in the Shuttle's cargo bay for the trip home.

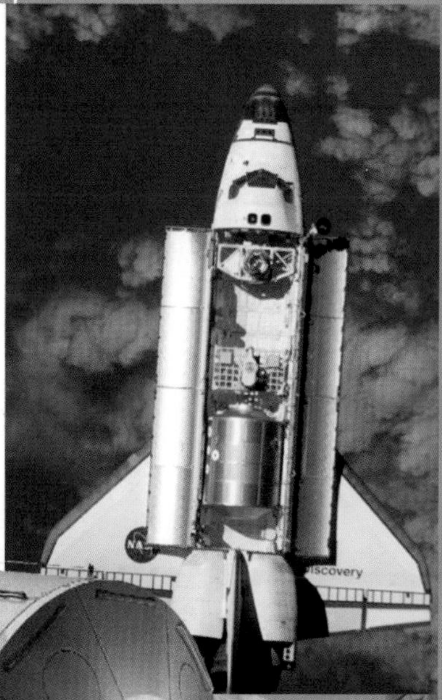

MPLM berthed at Node 1.

Stowage within MPLM.

MPLM interior during cargo transfers.

Length	6.6 m (21.7 ft)
Diameter	4.2 m (13.8 ft)
Mass (structure)	4,685 kg (10,329 lb)
Mass (payload)	9,400 kg (20,700 lb)
Racks	16, 5 active
Pressurized habitable volume	31 m^3 (1,095 ft^3)

JAXA H-II Transfer Vehicle (HTV)

Japan Aerospace Exploration Agency (JAXA)/ Mitsubishi Heavy Industries, Ltd.

The H-II Transfer Vehicle is an autonomous logistical resupply vehicle designed to berth to the International Space Station using the Space Station Remote Manipulation System (SSRMS). HTV offers the capability to carry logistics materials in both its internal pressurized carrier as well as in an unpressurized carrier for exterior placement. It is launched on the H-II unmanned launch vehicle and can carry dry cargo, gas and water, and propellant. After fresh cargo is unloaded at the ISS, the HTV is loaded with trash and waste products; after unberthing and deorbit, it is incinerated during reentry.

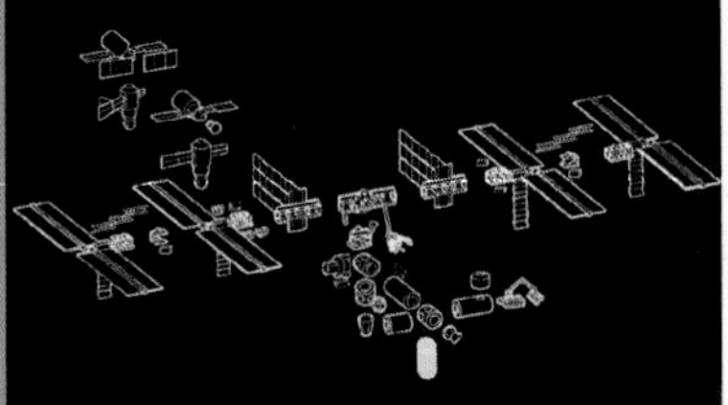

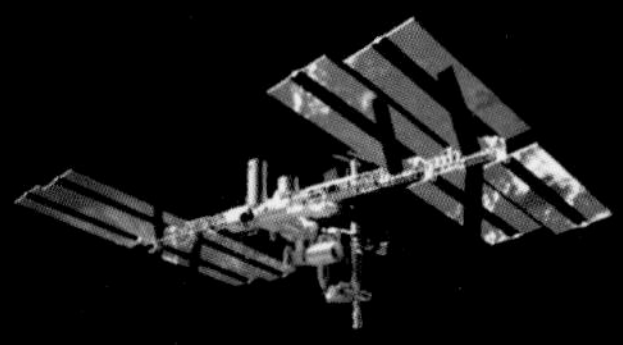

After rendezvous with the ISS, the HTV awaits grappling by the SSRMS.

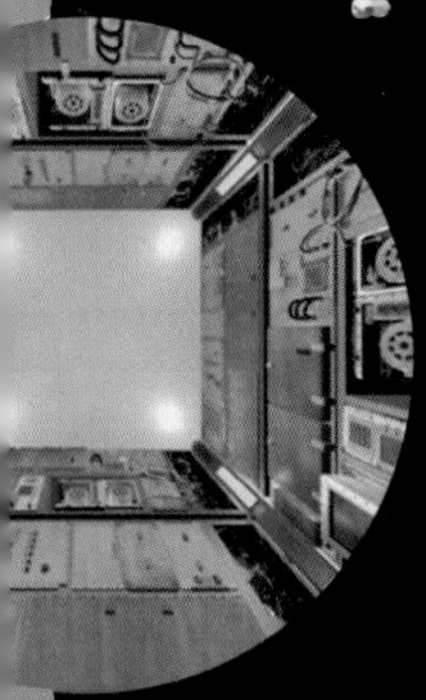

Interior View of HTV Pressurized Carrier.

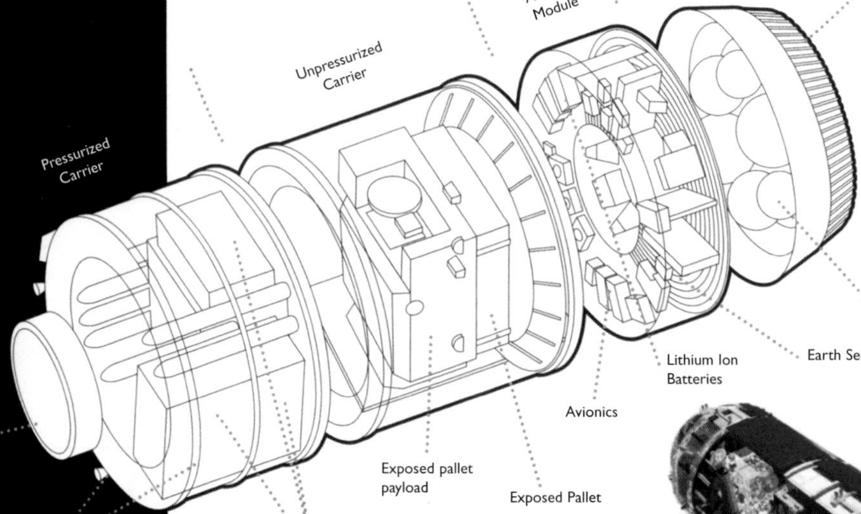

Length	9.2 m (30 ft)
Maximum diameter	4.4 m (14.4 ft)
Launch mass	16,500 kg (36,375 lb)
Cargo upload capacity	5,500 kg (12,125 lb)
Pressurized habitable volume	14 m^3 (495 ft^3)
Unpressurized volume	16 m^3 (565 ft^3)
Orbital life	6 mo

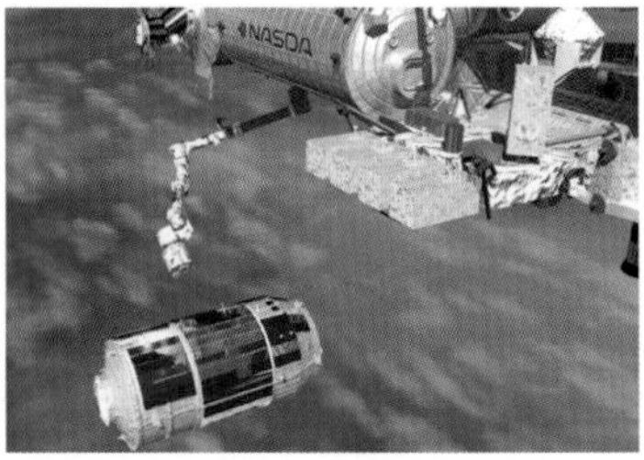

The HTV is berthed onto JEM by the Space Station RMS.

The HTV primary propulsion system performs rendezvous maneuvers.

Automated Transfer Vehicle (ATV)

European Space Agency (ESA)/European Aeronautic Defence and Space Co. (EADS)

The European Space Agency Automated Transfer Vehicle is an autonomous logistical resupply vehicle designed to dock to the International Space Station and provide the crew with dry cargo, atmospheric gas, water, and propellant. After the cargo is unloaded, the ATV is reloaded with trash and waste products, undocks, and is incinerated during reentry.

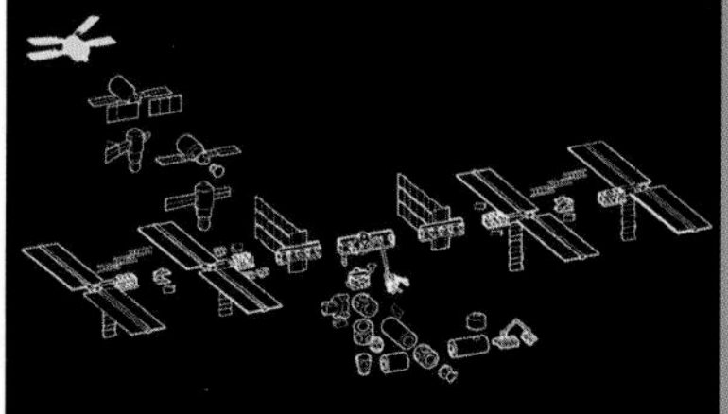

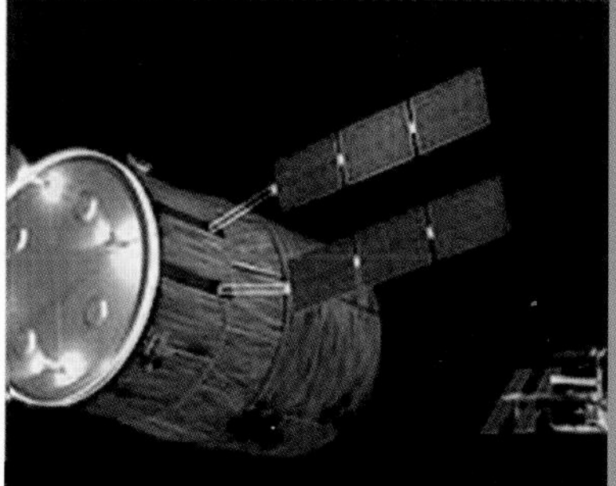

Artist's rendering shows the ATV approaching the ISS.

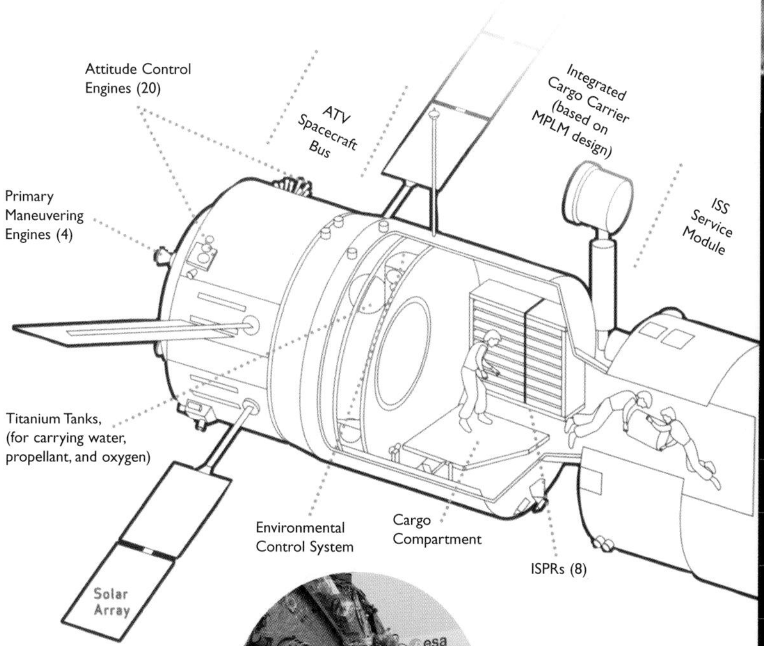

Russian-built probe and drogue docking system.

The ATV during manufacture.

Length	10.3 m (33.8 ft)
Maximum diameter	4.5 m (14.8 ft)
Span across solar arrays	22.3 m (73.2 ft)
Launch mass	20,750 kg (45,746 lb)
Cargo upload capacity	7,667 kg (16,903 lb)
Engine thrust	1,960 N (441 lbf)
Orbital life	6 mo
Cargo Load	
Dry cargo such as bags	5,500 kg (12,125 lb)
Water	840 kg (1,852 lb)
Air (O_2, N_2)	100 kg (220 lb)
Refueling propellant	860 kg (1,896 lb)
Reboost propellant	4,700 kg (10,360 lb)
Waste capacity	6,500 kg (14,330 lb)

Crew Exploration Vehicle (CEV)/Orion

NASA has initiated the development of the Orion Crew Exploration Vehicle (CEV). The first Orion flights are planned for 2012–2014 and will support the ISS.

The CEV approaches the ISS.

Commercial Orbital Transportation Services (COTS)

NASA is seeking commercial providers of launch and return logistics services to support the ISS after the Space Shuttle is retired. The first COTS demonstration missions are planned for 2010.

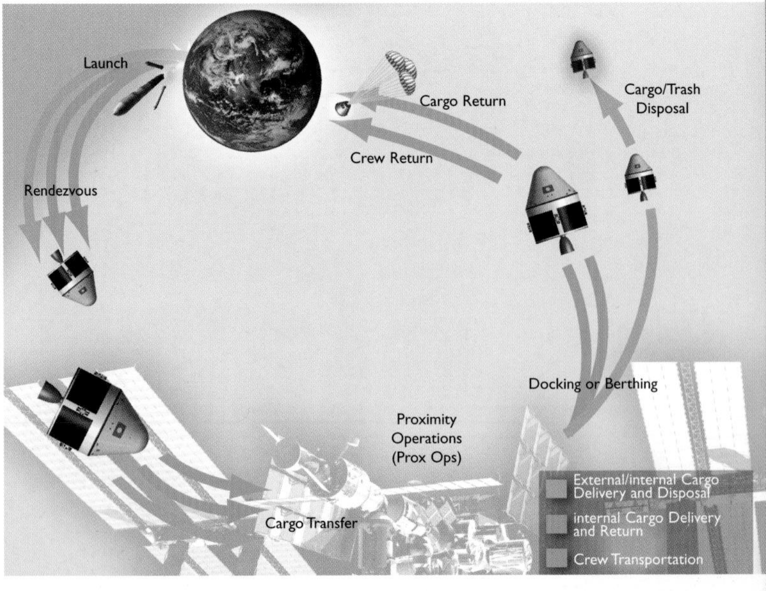

systems

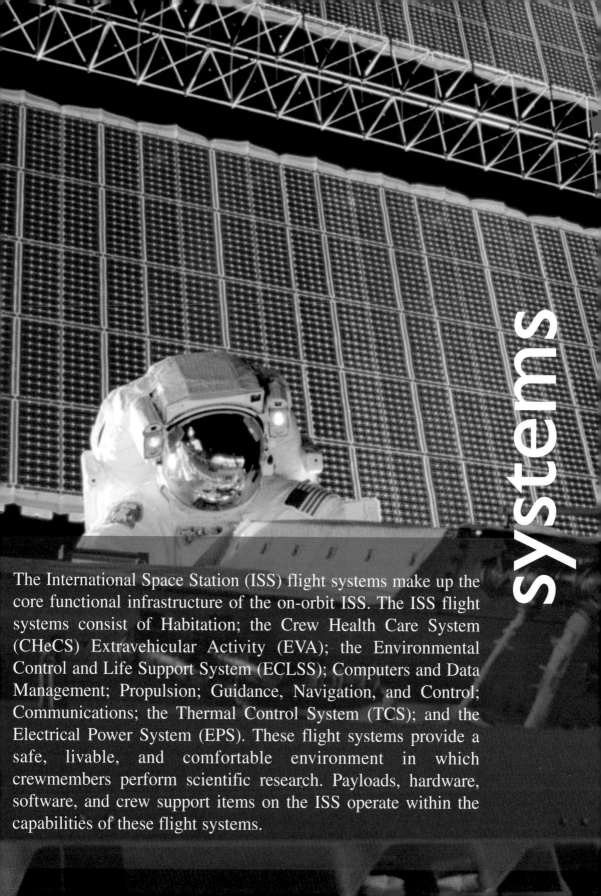

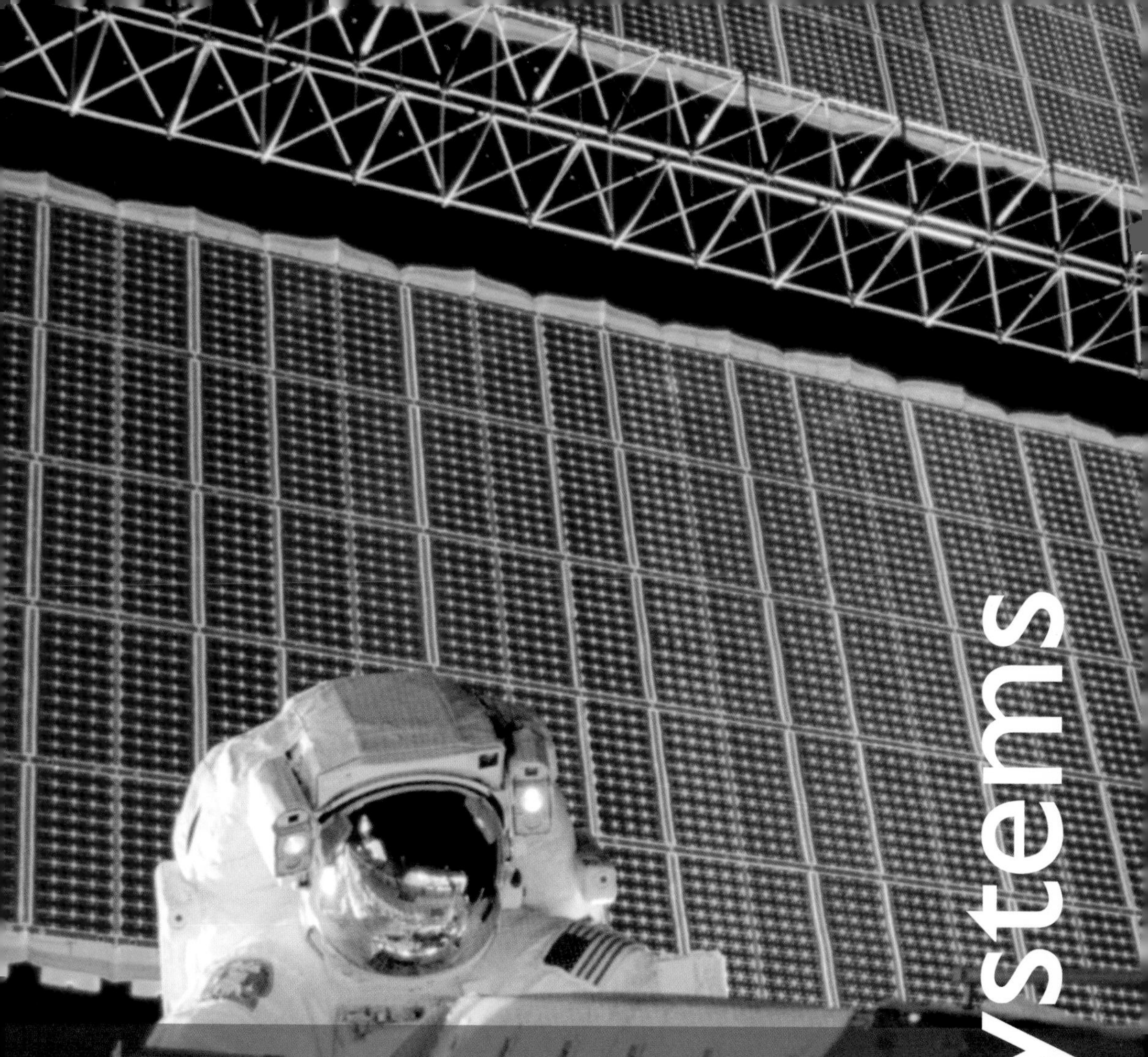

The International Space Station (ISS) flight systems make up the core functional infrastructure of the on-orbit ISS. The ISS flight systems consist of Habitation; the Crew Health Care System (CHeCS) Extravehicular Activity (EVA); the Environmental Control and Life Support System (ECLSS); Computers and Data Management; Propulsion; Guidance, Navigation, and Control; Communications; the Thermal Control System (TCS); and the Electrical Power System (EPS). These flight systems provide a safe, livable, and comfortable environment in which crewmembers perform scientific research. Payloads, hardware, software, and crew support items on the ISS operate within the capabilities of these flight systems.

Integrated Truss Assembly

The truss assemblies provide attachment points for the solar arrays, thermal control radiators, and external payloads. Truss assemblies also contain electrical and cooling utility lines, as well as the mobile transporter rails. The Integrated Truss Structure (ITS) is made up of 11 segments plus a separate component called Z1. These segments, which are shown in the figure, will be installed on the Station so that they extend symmetrically from the center of the ISS.

At full assembly, the truss reaches 108.5 meters (356 feet) in length across the extended solar arrays. ITS segments are labeled in accordance with their location. P stands for "port," S stands for "starboard," and Z stands for "Zenith."

Initially, through Stage 8A, the first truss segment, Zenith 1 (Z1), was attached to the Unity Node zenith berthing mechanism. Then truss segment P6 was mounted on top of Z1 and its solar arrays and radiator panels deployed to support the early ISS. Subsequently, S0 was mounted on top of the U.S. Lab Destiny, and the horizontal truss members P1 and S1 were then attached to S0. As the remaining members of the truss are added, P6 will be removed from its location on Z1 and moved to the outer end of the port side.

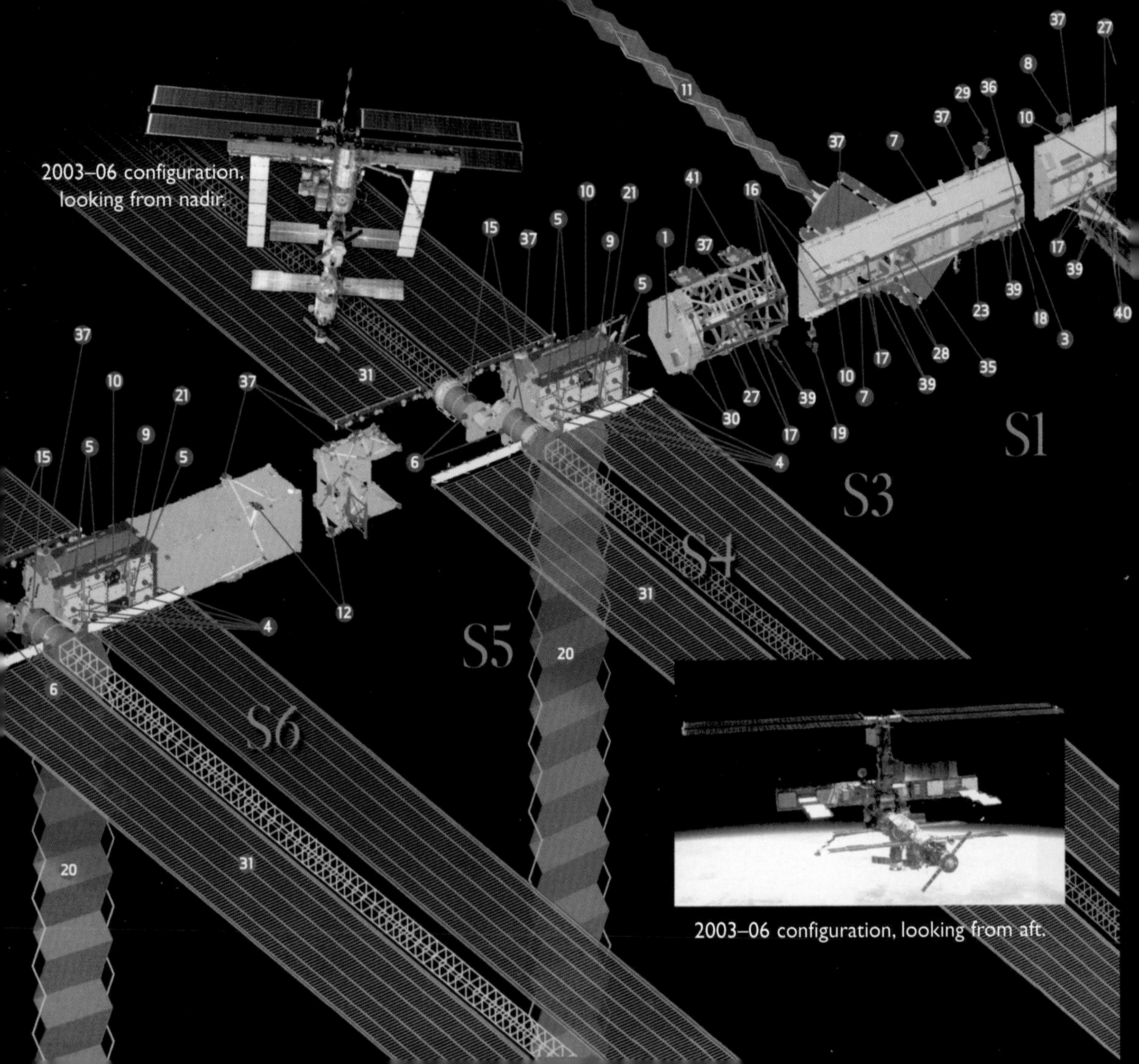

2003–06 configuration, looking from nadir.

2003–06 configuration, looking from aft.

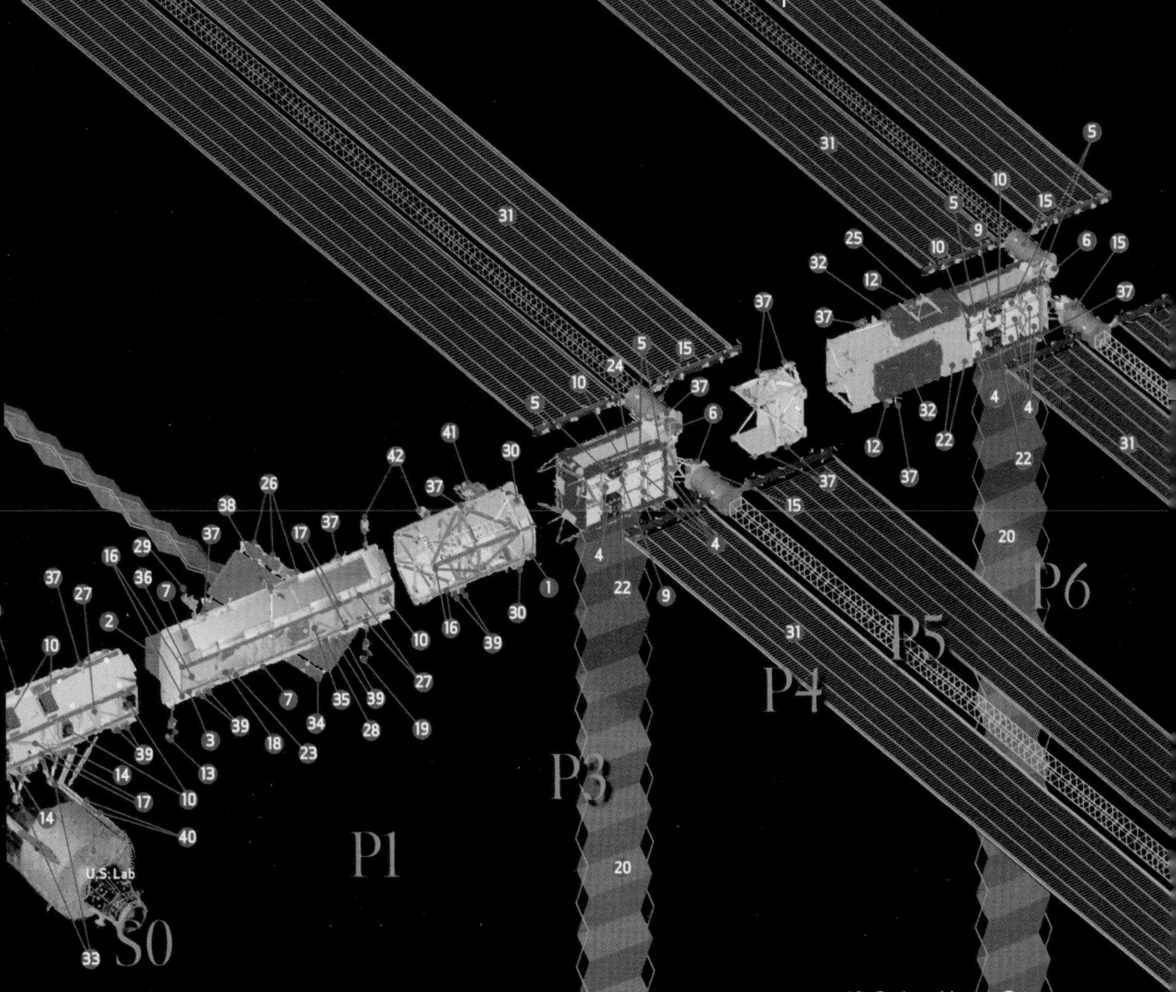

1 Solar Array Alpha Rotary Joint
2 Ammonia Tank Assembly
3 Assembly Contingency Baseband Signal Processor
4 Batteries
5 Battery Charge Discharge Unit
6 Beta Gimbal Assemblies
7 Cable Trays
8 Charged Particle Directional Spectrometer
9 Direct Current Switching Unit (DCSU)
10 DC-to-DC Converter Unit (DDCU)
11 Deployed Thermal System Radiator
12 Grapple Fixture
13 Inboard Lower Camera
14 Main Bus Switching Units
15 Mast Storage Canister
16 Mobile Transporter Rails
17 Multiplexer/De-Multiplexers
18 Nitrogen Tank Assembly (interior to truss)
19 Outboard Lower Camera
20 Photovoltaic Radiator
21 Pump Flow Control Assembly
22 Pump Flow Control Subassembly
23 Pump Module
24 PVR Controller Unit
25 PVR Grapple Fixture Bar
26 Radiator Beam Valve Module
27 Remote Power Control Modules
28 Rotary Joint Motor Controller
29 S-Band Antenna
30 Solar Array Alpha Rotary Joint Drive Lock Assembly
31 Solar Array Wing
32 Stowed Photovoltaic Radiator
33 Struts
34 Thermal Control System Radiator Beam
35 Thermal Radiator Rotary Joint with Flex Hose Rotary Coupler
36 Transponder
37 Trunnion
38 UHF Antenna
39 Umbilical Mechanism Assemblies
40 Umbilicals
41 Unpressurized Cargo Carrier Attachment
42 Wireless Video System Antenna

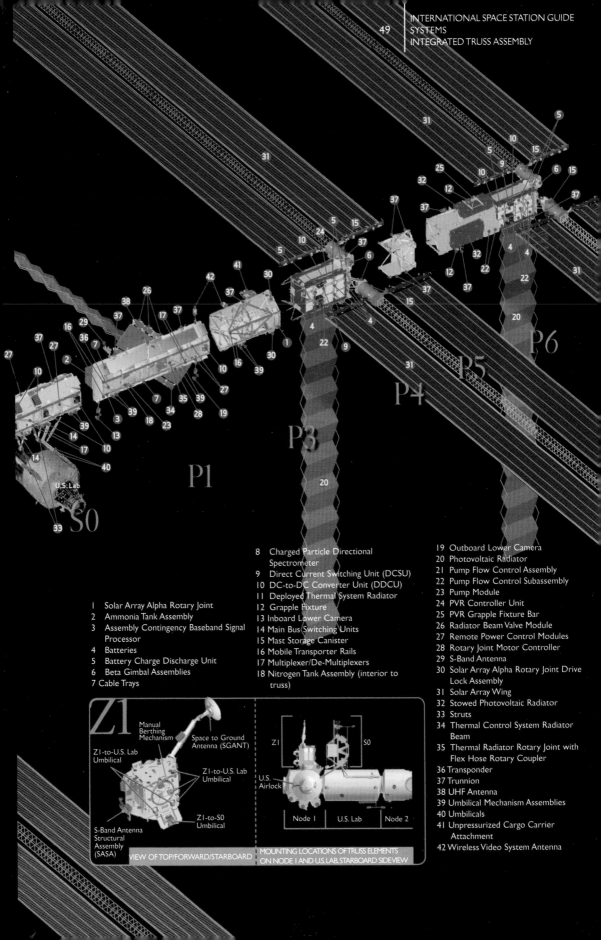

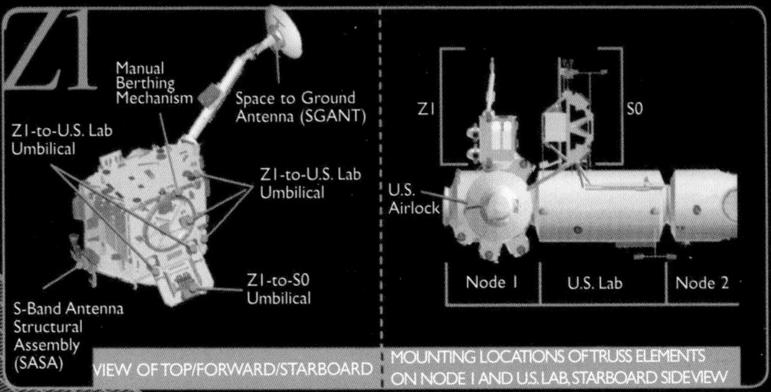

Habitation

The habitable elements of the International Space Station are mainly a series of cylindrical modules. Many of the primary accommodations, including the waste management compartment and toilet, the galley, individual crew sleep compartments, and some of the exercise facilities, are in the Service Module (SM). A third sleep compartment is located in the U.S. Lab, and additional exercise equipment is in the U.S. Lab and the Node. Additional habitation capabilities for a crew of six will be provided prior to completion of ISS assembly.

Haircut in SM.

Shaving in SM.

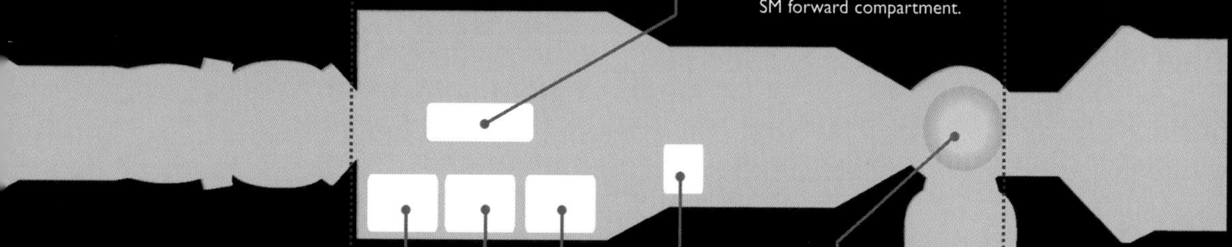

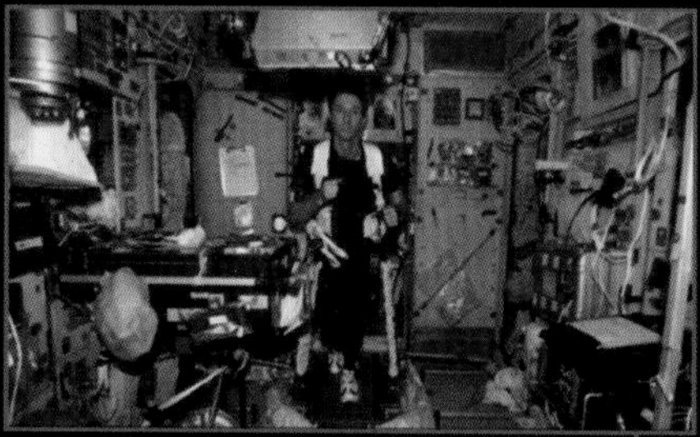
SM mid compartment and treadmill.

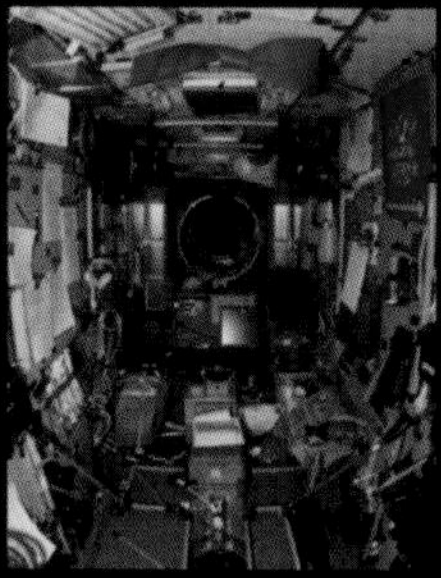
SM forward compartment.

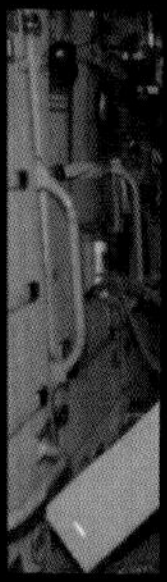

Russian water containers.

SM Sleep Compartment

Remote Docking Control Station

SM Transfer Compartment

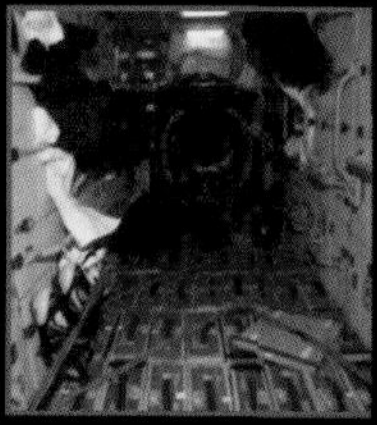
Stowed Food Trays in FGB

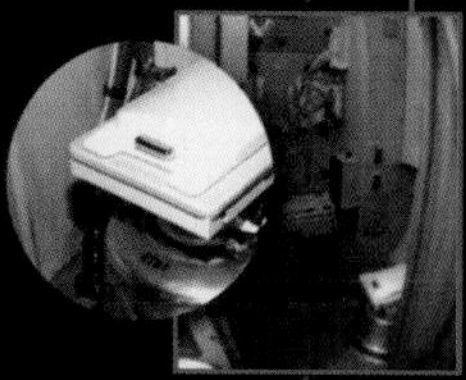
Toilet in Waste Management Compartment

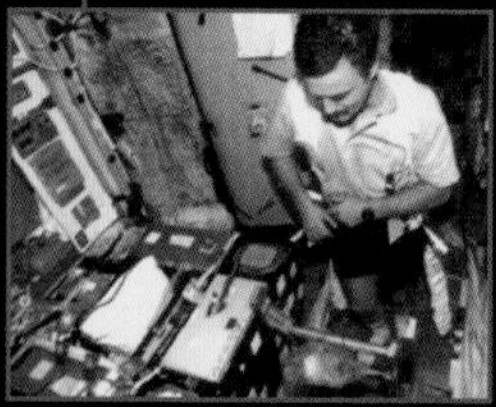
Crewmembers Exercise on SM Treadmill

Crewmembers with Orlan Suits in Pirs

Preparing meal in galley.

Playing keyboard in U.S. Lab.

fgb

Stowage container in FGB.

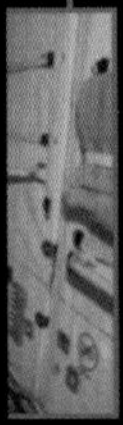

FGB Corridor and Stowage

node/airlock

U.S./Joint Airlock

Node Passageway

Stowage in Node 1

u.s. lab

U.S. Lab Computer Workstation

U.S. Lab Temporary Sleep Station (TSS)

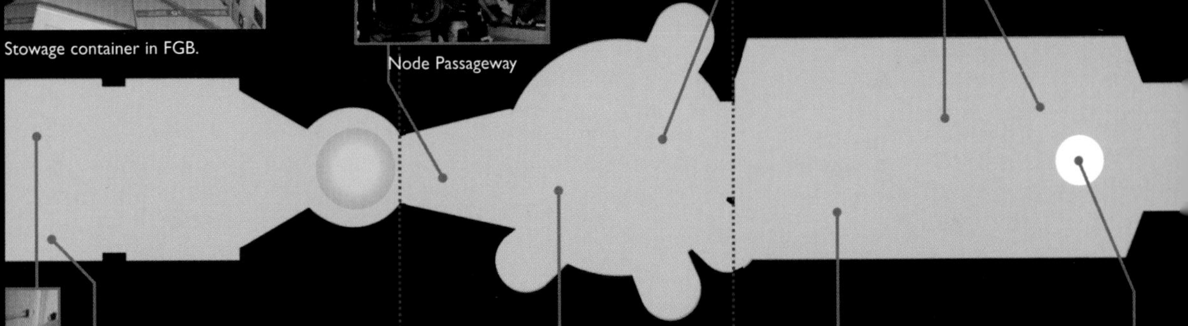

Microgravity Science Glovebox in U.S. Lab

U.S. Lab Window

soyuz ... service module

Water Sampling and Analysis

Treadmill Vibration Isolation System (TVIS)

Blood Sample Reflotron

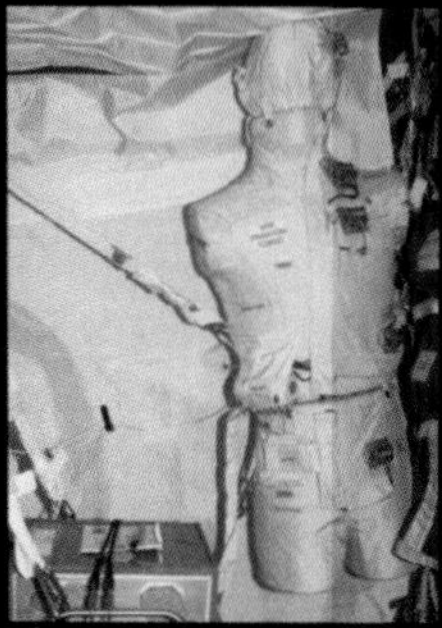

Saliva Sample Kit

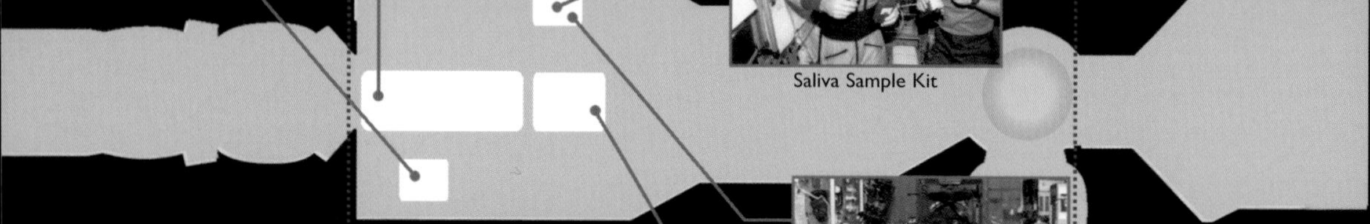

CardioCog

Countermeasures System (CMS)—The CMS provides the equipment and protocols for the performance of daily and alternative regimens (e.g., exercise) to mitigate the deconditioning effects of living in a microgravity environment. The CMS also monitors crewmembers during exercise regimens, reduces vibrations during the performance of these regimens, and makes periodic fitness evaluations possible.

Environmental Health System (EHS)—The EHS monitors the atmosphere for gaseous contaminants (i.e., from nonmetallic materials off-gassing, combustion products, and propellants), microbial contaminants (i.e., from crewmembers and Station activities), water quality, acoustics, and radiation levels.

Health Maintenance System (HMS)—The HMS provides in-flight life support and resuscitation, medical care, and health monitoring capabilities.

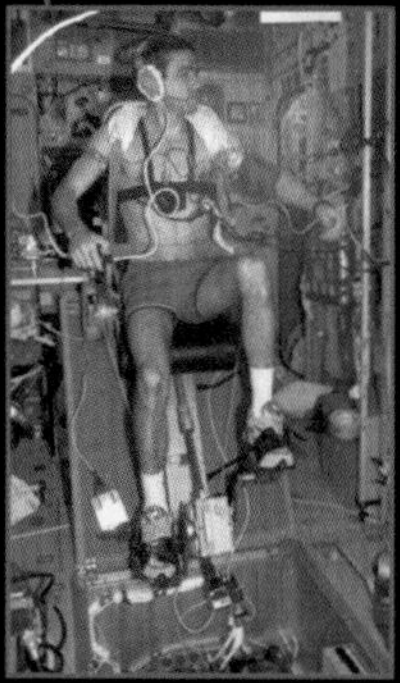
Velo-Ergometer

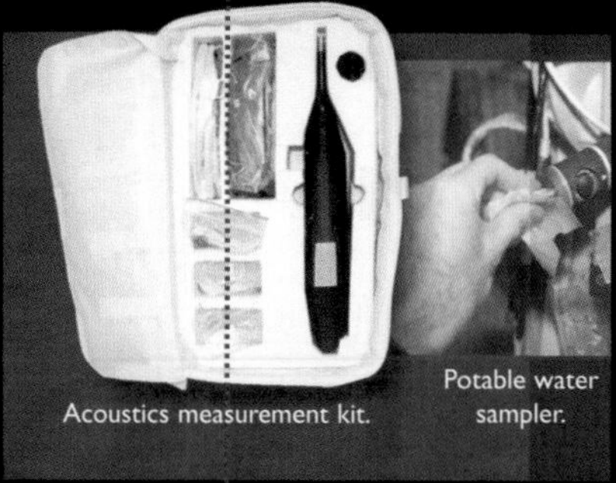
Acoustics measurement kit.

Potable water sampler.

Leroy Chiao uses RED.

Crew uses medical restraint and defibrillator.

fgb

node/airlock

u.s. lab

Bonner Ball Neutron Particle Detector and Phantom Torso for radiation measurement experiments.

Resistive Exercise Device (RED)

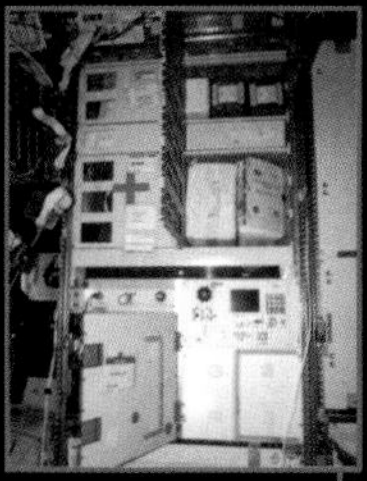

CHeCS Rack

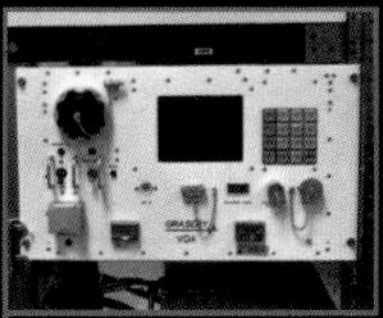

Volatile Organics Analyzer (VOA)

Water Samples (taken for ground analysis of contamination)

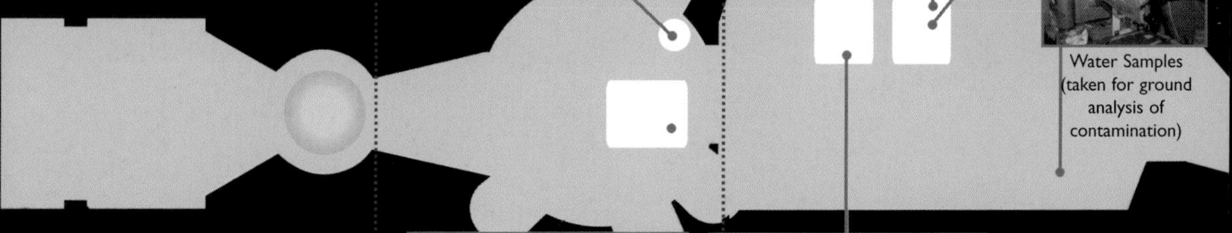

Microbial Surface Sampling

Cycle Ergometer with Vibration Isolation System (CEVIS)

From left to right: Intravehicular Charged Particle Directional Spectrometer (IV-CPDS) (gold box) and Tissue Equivalent Proportional Counter (TEPC) detector (gold cylinder).

Atmosphere Grab Sampler Container.

Crew Medical Restraint System (CMRS).

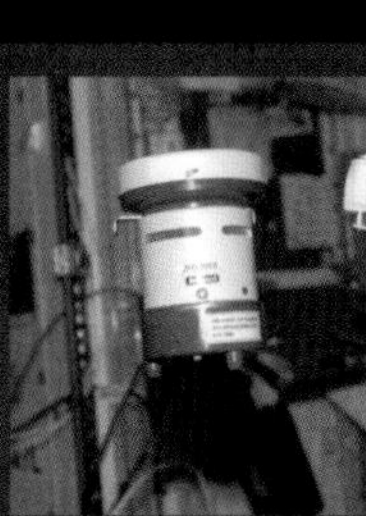

Microbial air sampler.

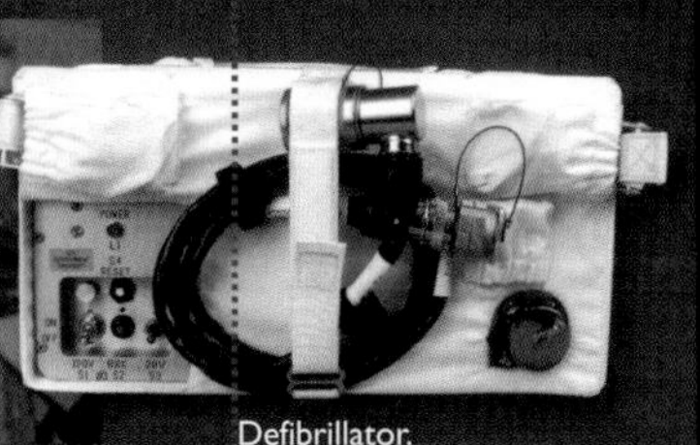

Defibrillator.

Environmental Control and Life Support System (ECLSS)

Earth's natural life-support system provides the air we breathe, the water we drink, and other conditions that support life. For people to live in space, however, these functions must be performed by artificial means. The ECLSS includes compact and powerful systems that provide the crew with a comfortable environment in which to live and work.

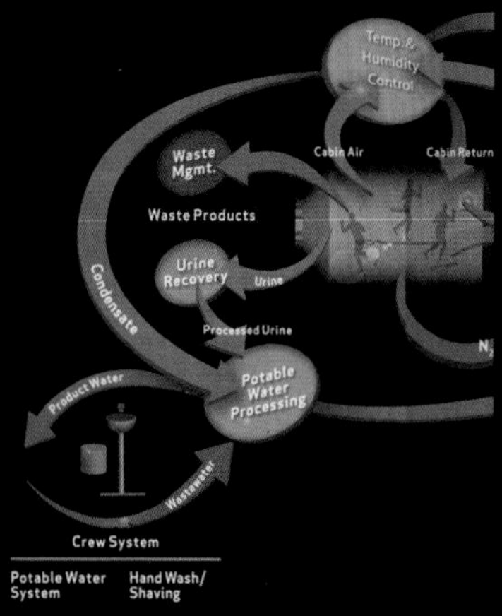

progress · service module · fgb

elektron (produces oxygen from water through electrolysis; vents hydrogen out of the Station)

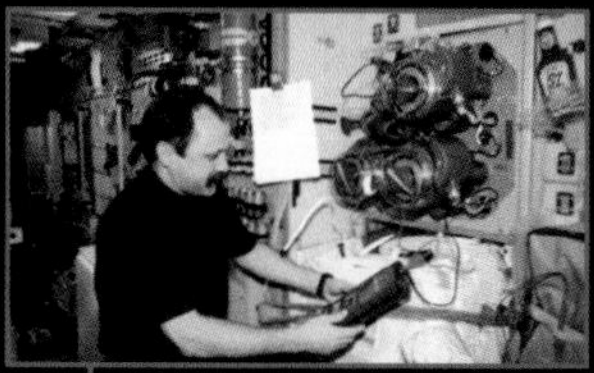

Vozdukh (absorbs carbon dioxide from crew)

Russian EDVs used to store and transport water.

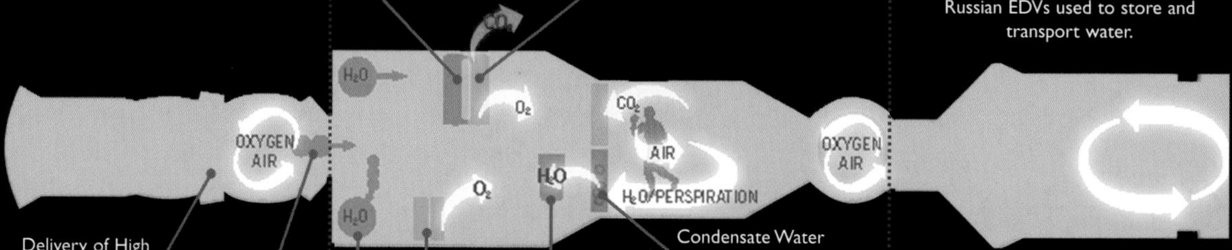

Delivery of High Pressure Oxygen and Air on Progress

Water Delivery from Progress

Freshwater Storage Tanks

Condensate Water Processor (condenses water vapor from air)

Russian EDVs (used to store reclaimed water)

SM gas analyzer.

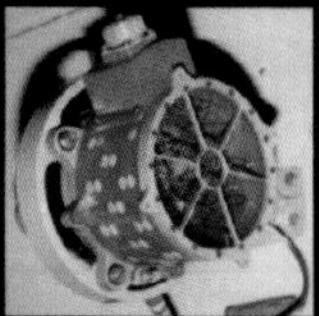

Airflow ventilation fan.

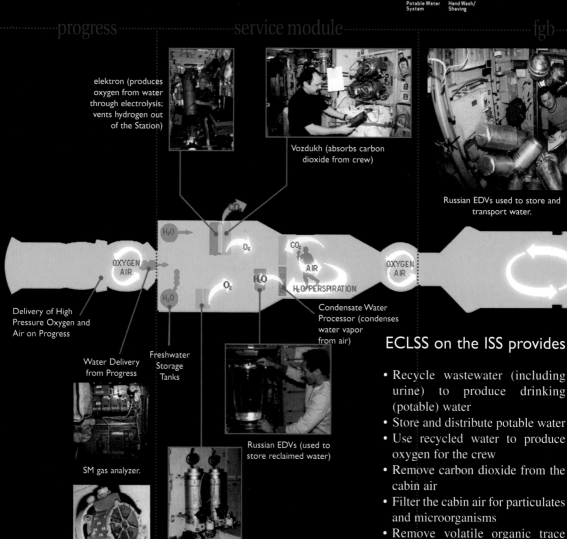

Solid Fuel Oxygen Generator (SFOG, burns candles to produce oxygen, backup system)

ECLSS on the ISS provides

- Recycle wastewater (including urine) to produce drinking (potable) water
- Store and distribute potable water
- Use recycled water to produce oxygen for the crew
- Remove carbon dioxide from the cabin air
- Filter the cabin air for particulates and microorganisms
- Remove volatile organic trace gases from the cabin air
- Monitor and control cabin air partial pressures of nitrogen, oxygen, carbon dioxide, methane, hydrogen, and water vapor

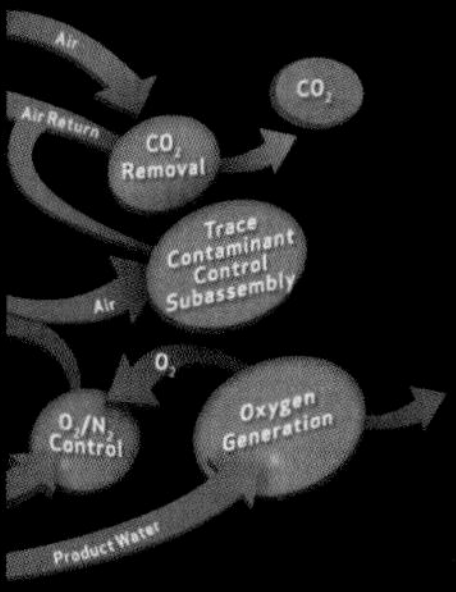

Regenerative environmental control life support in the U.S. segment of the ISS.

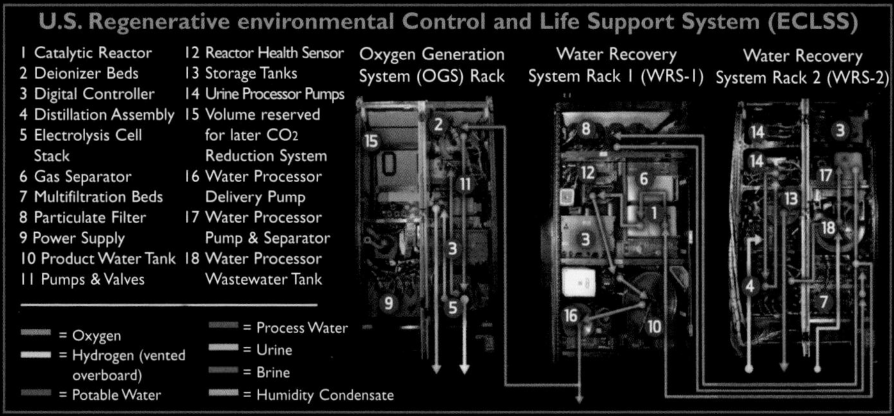

Lithium Hydroxide (LiOH) cartridge used for eliminating CO2 from air, backup system.

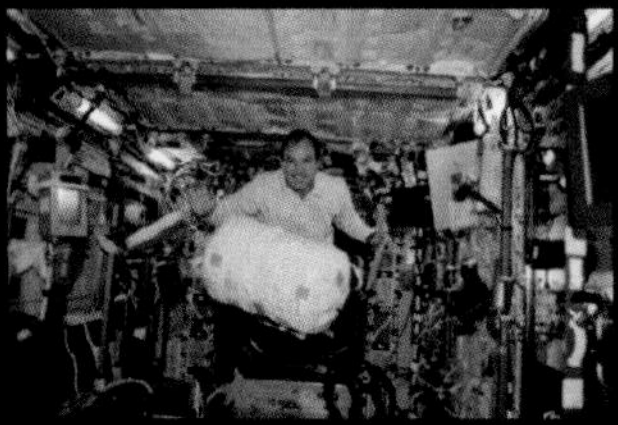

Contingency Water Container (CWC) bag.

Lithium Hydroxide Cartridges (absorb carbon dioxide)

CWC Bags (used by astronauts to carry water from the Shuttle to the ISS)

Fuel Cells (make electricity and water from oxygen and hydrogen)

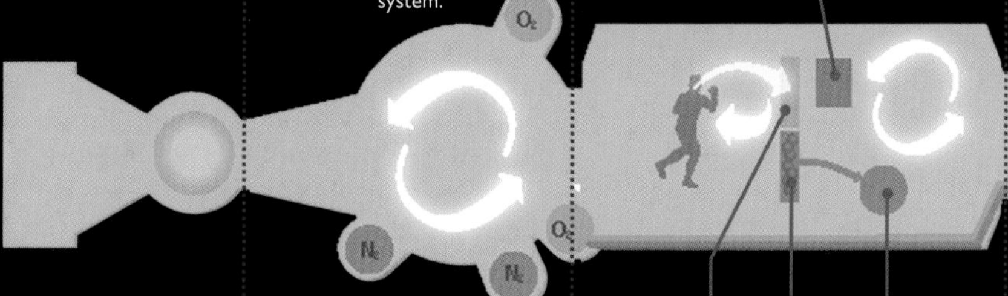

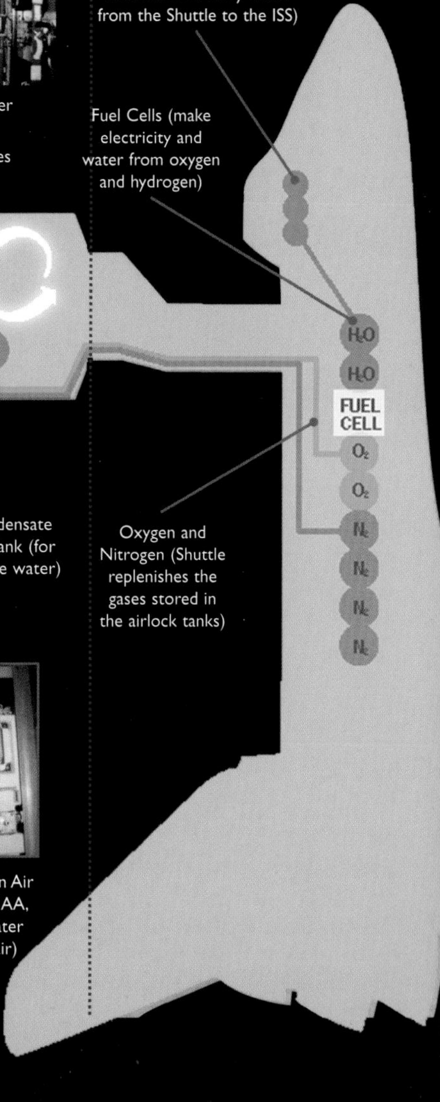

the following functions:

- Maintain total cabin pressure
- Detect and suppress fire
- Maintain cabin temperature and humidity levels
- Distribute cabin air between ISS modules (ventilation)

In the future, a new U.S. Regenerative Environmental Control and Life Support System will take additional steps toward closing the water cycle; it will take humidity condensate from the cabin air and urine from the crew and convert these into drinking water, oxygen for breathing, and hydrogen.

Carbon Dioxide Removal Assembly (CDRA, adsorbs carbon dioxide from crew)

Lab Condensate Storage Tank (for condensate water)

Oxygen and Nitrogen (Shuttle replenishes the gases stored in the airlock tanks)

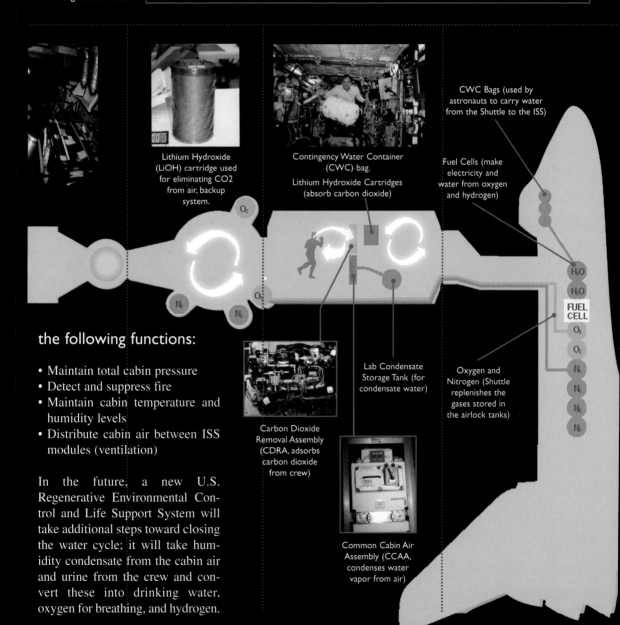

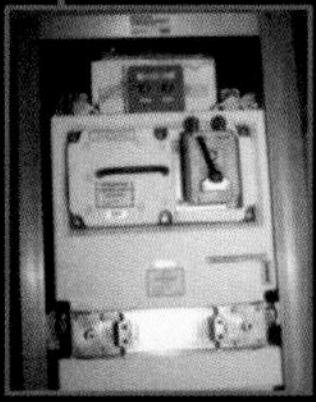

Common Cabin Air Assembly (CCAA, condenses water vapor from air)

soyuz

service module

Laptop (in SM crew quarters)

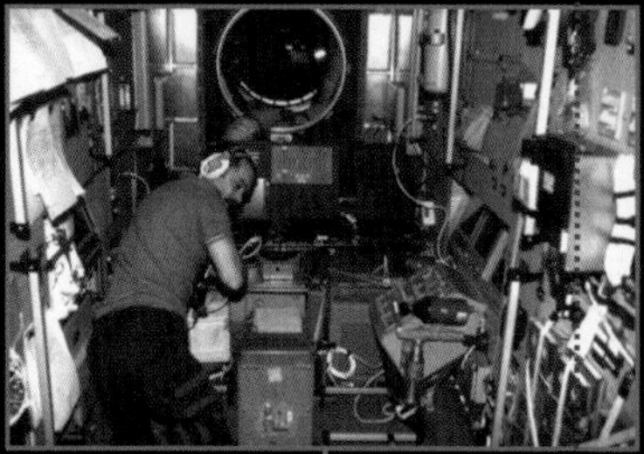
Primary Command Workstation in SM

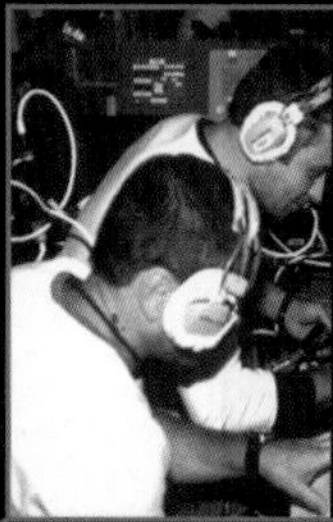
Crew uses Progress Remote Control workstation in SM

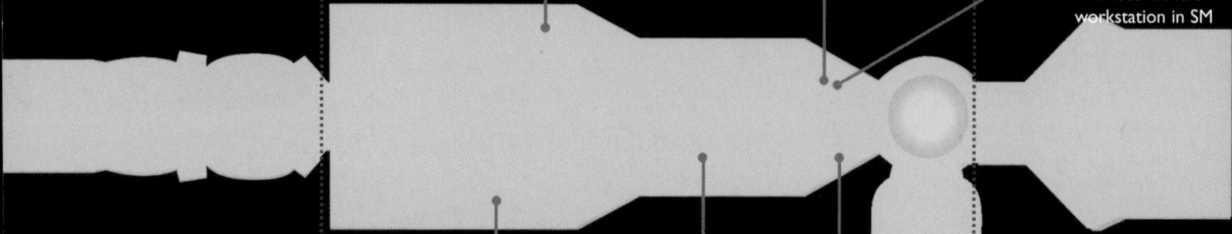

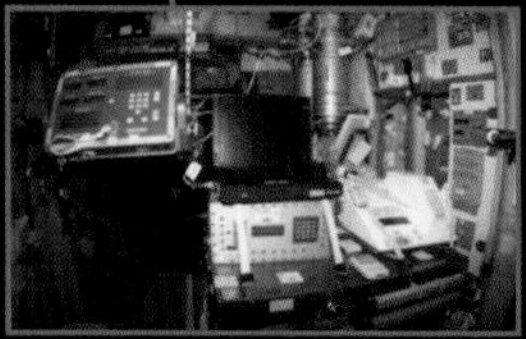
Laptop and TVIS Control (located near galley)

TORU Remote Progress Docking Workstation

Russian Segment Workstations

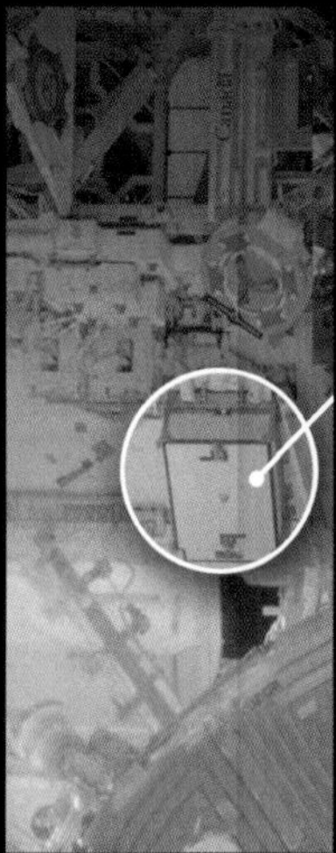

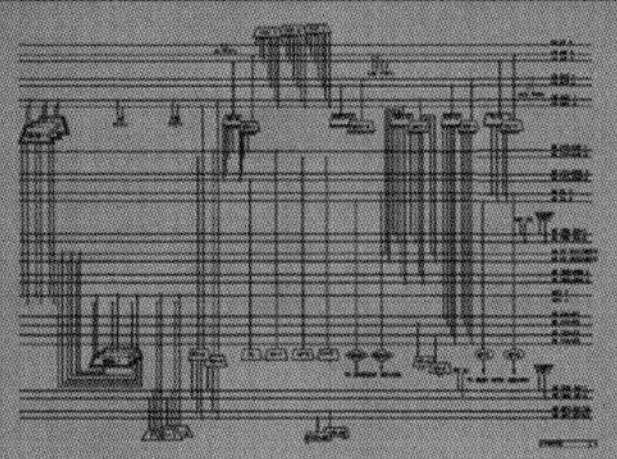

Data bus architecture consists of

- 100+ MIL-STD-1553B data buses,
- 60+ computers into which software can be loaded as necessary,
- 1,200+ remote terminals,
- 190 payload remote terminals,
- 600+ international partner and firmware controller devices, and
- 90+ unique types of remote devices.

SSRMS Control and Robotics Workstations

fgb

node/airlock

u.s. lab

Maneuvering Truss Segments into Place at SSRMS Workstation

Multiplexer/Demultiplexer (computer)

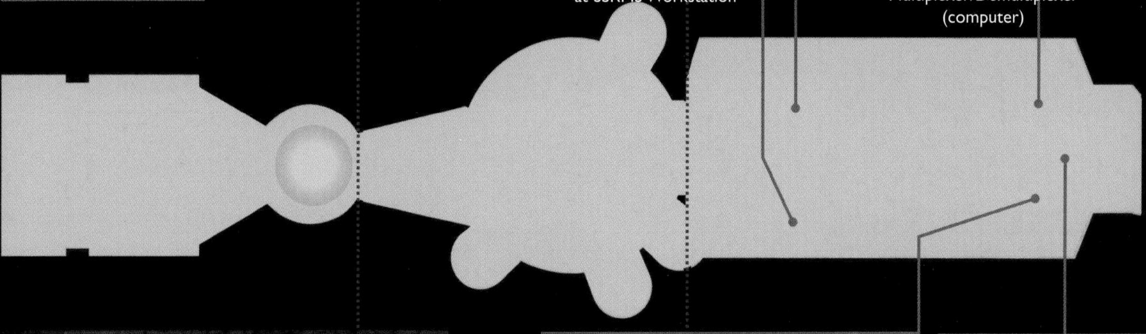

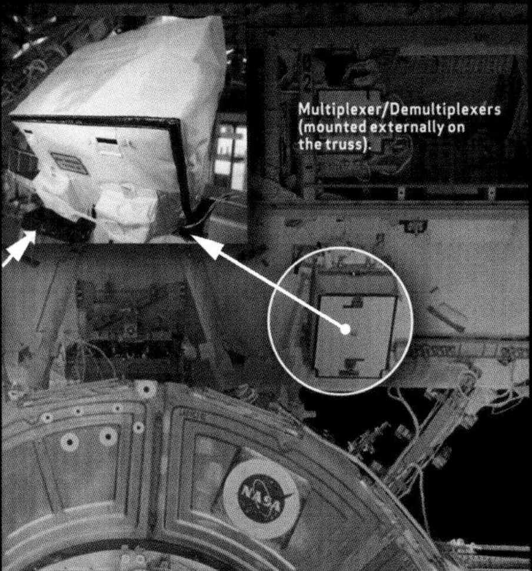

Multiplexer/Demultiplexers (mounted externally on the truss).

Human Research Facility Workstation

Multiplexer/Demultiplexer Mass Memory Unit (MMU) Processor Data Cards in U.S. Lab

progress

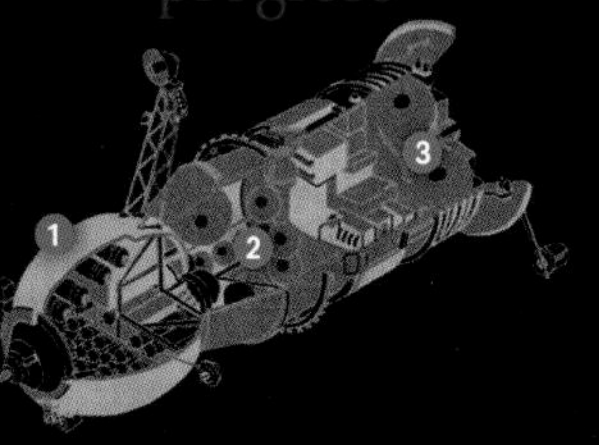

1 Progress Cargo Module
2 Propellant Resupply Tanks
3 Progress Propulsion System

service module

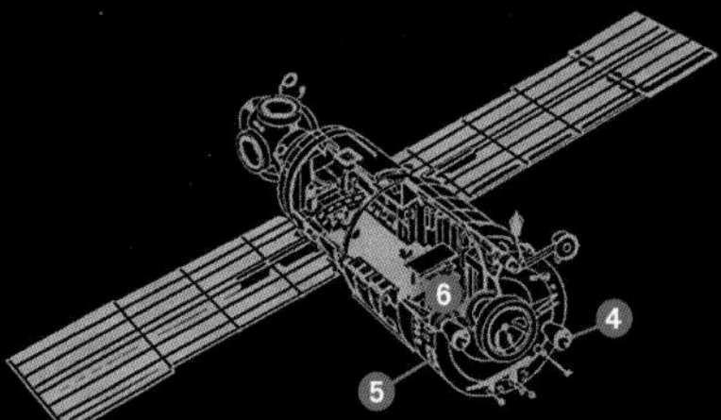

4 Main Engines (2)
5 Attitude Control Engines (32)
6 Propellant Tanks (4)

fgb

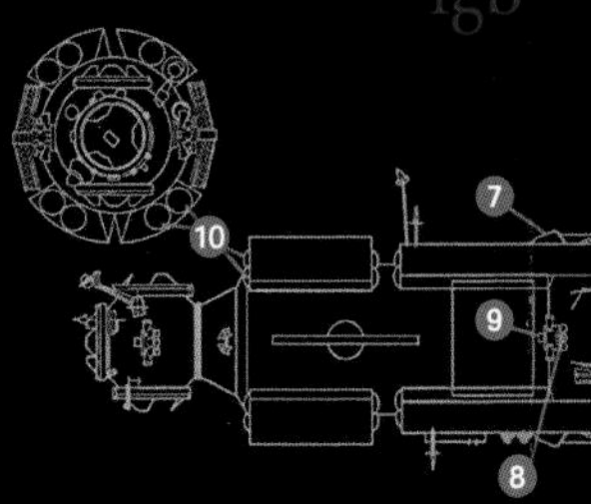

7 Correction and Docking Engines (2)
8 Docking and Stabilization Engines (24)
9 Accurate Stabilization Engines (16)
10 Propellant Tanks (16)

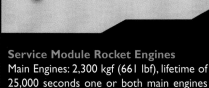

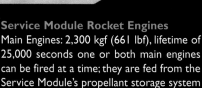

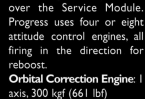

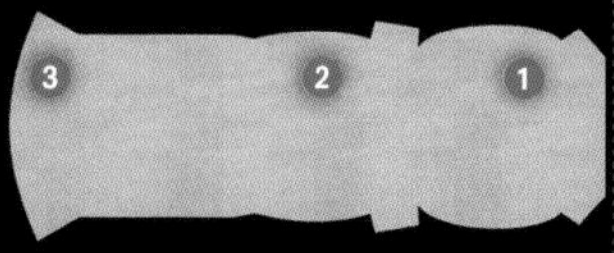

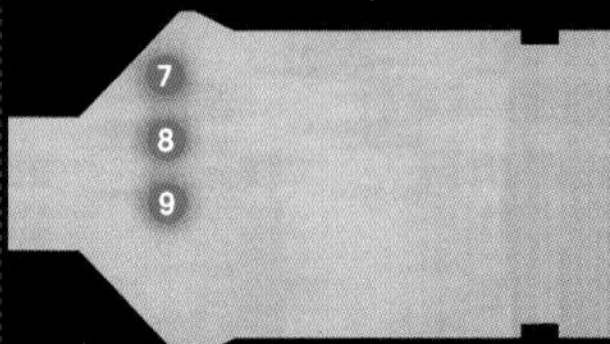

Progress Rocket Engines

Progress is used for propellant resupply and for performing reboosts. For the latter, Progress is preferred over the Service Module. Progress uses four or eight attitude control engines, all firing in the direction for reboost.

Orbital Correction Engine: 1 axis, 300 kgf (661 lbf)

Attitude Control Engines: 28 multidirectional, 13.3 kgf (29.3 lbf)

Service Module Rocket Engines

Main Engines: 2,300 kgf (661 lbf), lifetime of 25,000 seconds one or both main engines can be fired at a time; they are fed from the Service Module's propellant storage system

Attitude Control Engines: 32 multidirectional, 13.3 kgf (29.3 lbf); attitude control engines can accept propellant fed from the Service Module, the attached Progress, or the FGB propellant tanks

Service Module Propellant Storage

Two pairs of 200-L (52.8-gal) propellant tanks (two nitrogen tetroxide N_2O_4 and two unsymmetrical dimethyl hydrazine [UDMH]) provide a total of 860 kg (1,896 lb) of usable propellant. The propulsion system rocket engines use the hypergolic reaction of UDMH and N2O4. The Module employs a pressurization system using N2 to manage the flow of propellants to the engines.

FGB Rocket Engines

FGB engines are deactivated once the Service Module is in use.

Correction and Docking Engines: 2 axis, 417 kgf (919 lbf)

Docking and Stabilization Engines: 24 multidirectional, 40 kgf (88 lbf)

Accurate Stabilization Engines: 16 multidirectional, 1.3 kgf (2.86 lbf)

FGB Propellant Storage

There are two types of propellant tanks in the Russian propulsion system: bellows tanks (SM, FGB), able both to receive and to deliver propellant, and diaphragm tanks (Progress), able only to deliver fuel.

Sixteen tanks provide 5,760 kg (12,698 lb) of N2O4 and UDMH storage: eight long tanks, each holding 400 L (105.6 gal), and eight short tanks, each holding 330 L (87.17 gal).

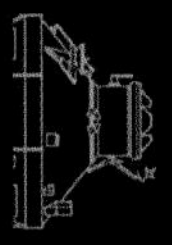

The ISS orbits Earth at an altitude that ranges from 370 to 460 kilometers (230 to 286 miles) and a speed of 28,000 kilometers per hour (17,500 miles per hour). Owing to atmospheric drag, the ISS is constantly slowed. Therefore, the ISS must be reboosted periodically in order to maintain its altitude. The ISS must sometimes be maneuvered in order to avoid debris in orbit. Furthermore, the ISS attitude control and maneuvering system can be used to assist in rendezvous and dockings with visiting vehicles, although that capability is not usually required.

Although the ISS typically relies upon large gyrodynes, which utilize electrical power, to control its orientation (see “Guidance, Navigation, and Control”), when force that is beyond the production capability of the gyrodynes is required, rocket engines provide propulsion for reorientation.

Rocket engines are located on the Service Module, as well as on the Progress, Soyuz, and Space Shuttle spacecraft.

The Service Module provides 32 13.3-kilograms force (29.3-pounds force) attitude control engines. The engines are combined into two groups of 16 engines each, taking care of pitch, yaw, and roll control. Each Progress provides 24 engines similar to those on the Service Module. When a Progress is docked at the aft Service Module port, these engines can be used for pitch and yaw control. When the Progress is docked at the Russian Docking Module, the Progress engines can be used for roll control.

Besides being a resupply vehicle, the Progress provides a primary method for reboosting the ISS. Eight 13.3-kilograms force (29.3-pounds force) Progress engines can be used for reboosting. Engines on the Service Module, Soyuz vehicles, and Space Shuttle can also be used. The Progress can also be used to resupply propellants stored in the FGB that are used in the Service Module engines. The ESA ATV and JAXA HTV will also provide propulsion and reboost capability.

Extravehicular Activity (EVA)

To date, there have been more than 69 EVAs (operations outside of the ISS pressurized modules) from the ISS totaling some 400 hours. Approximately 124 spacewalks, totaling over 900 hours,dedicated to assembly and maintenance of the Station will have been accomplished by Assembly Complete. Most of these EVAs have been for assembly tasks,but many were for maintenance, repairs, and science. These tasks were conducted from three different airlocks —the Shuttle Airlock, the U.S. Quest Airlock, and the Russian Pirs. Early in the program, an EVA was conducted from the Service Module Transfer Compartment. EVAs are conducted using two different spacesuit designs, the U.S. Extravehicular Mobility Unit (EMU) and the Russian Orlan.

The operational lessons of the ISS in the areas of EVA suit maintainability, training, and EVA support may prove critical for long-duration crewed missions that venture even further from Earth.

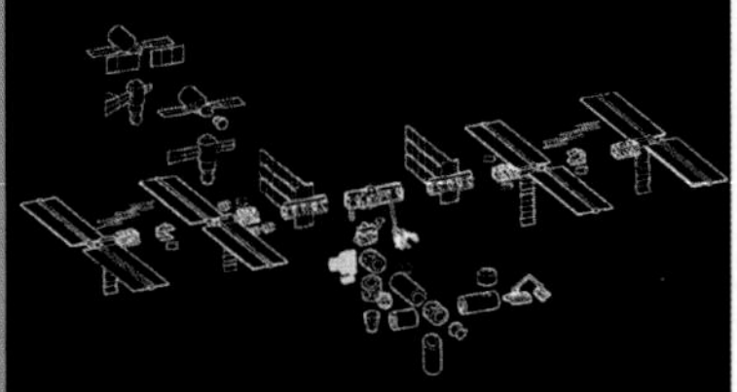

U.S./Joint Airlock (Quest)

NASA/Boeing

The Quest airlock provides the capability for extravehicular activity (EVA) using the U.S. Extravehicular Mobility Unit (EMU). The airlock consists of two compartments: the Equipment Lock, which provides the systems and volume for suit maintenance and refurbishment, and the Crew Lock, which provides the actual exit for performing EVAs. The Crew Lock design is based on the Space Shuttle's airlock design.

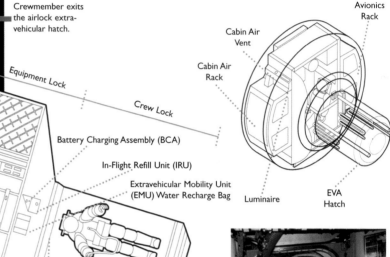

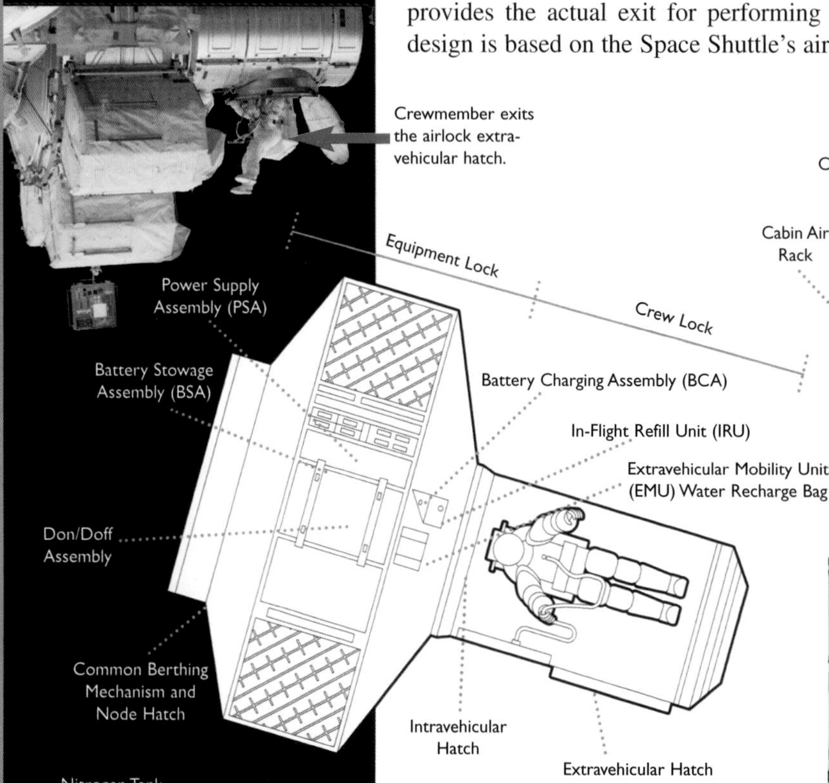

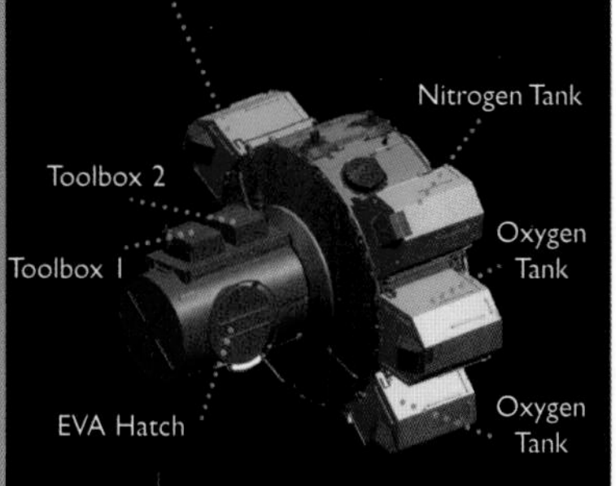

Length	5.5 m (18 ft)
Width	4.0 m (13.1 ft)
Mass	9,923 kg (21,877 lb)
Launch date	July 2001, on STS-104, ISS flight 7A. The Shuttle berthed to the starboard side of Node 1.

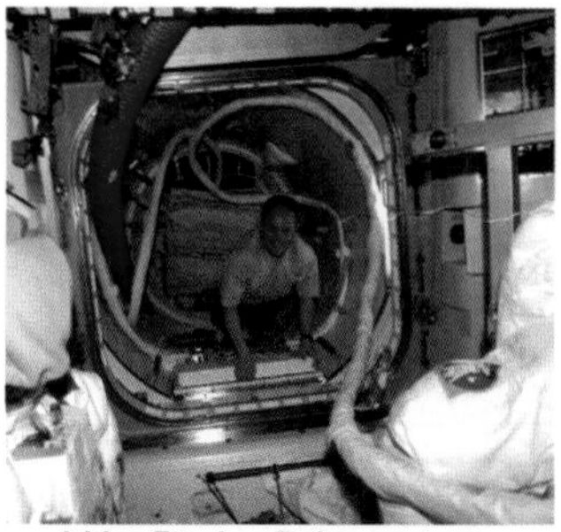

Mike Fincke, flight engineer on Expedition 9, inside Quest's Equipment Lock.

Space Shuttle mission STS-104 berths Quest to the starboard side of Node 1 in July 2001.

Airlock in preparation for launch in the Space Station Processing Facility at Kennedy Space Center.

Extravehicular Mobility Unit (EMU)

NASA/Hamilton Sundstrand/ILC Dover

The EMU provides a crewmember with life support and an enclosure that enables EVA. The unit consists of two major subsystems: the Life Support Subsystem (LSS) and the Space Suit Assembly (SSA). The EMU provides atmospheric containment, thermal insulation, cooling, solar radiation protection, and micrometeoroid/orbital debris (MMOD) protection.

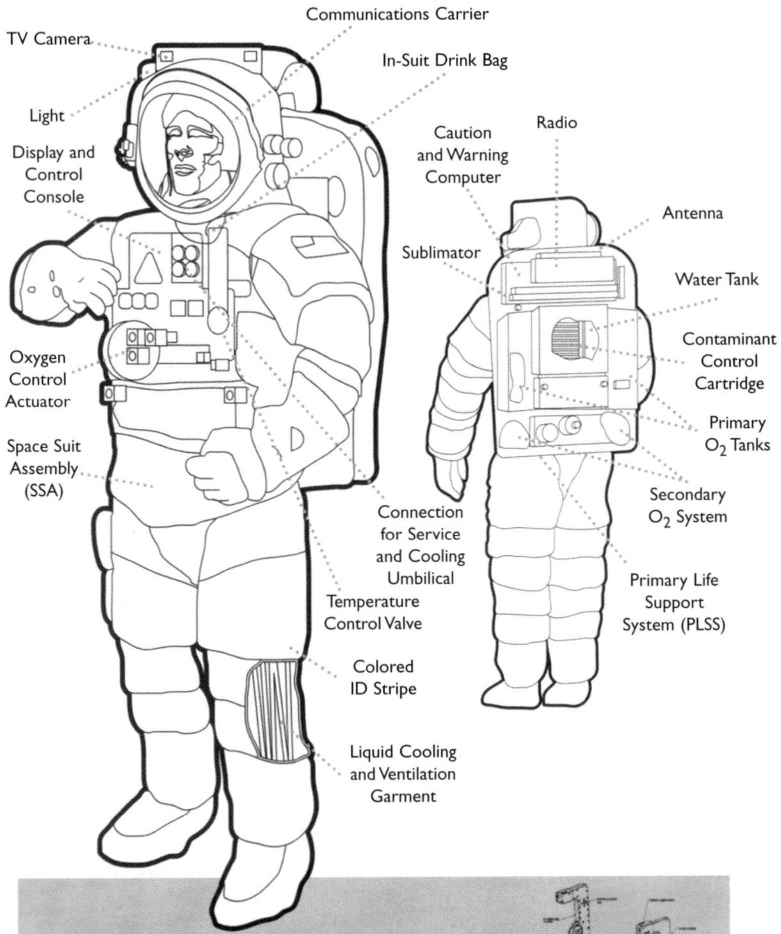

Suit Layers

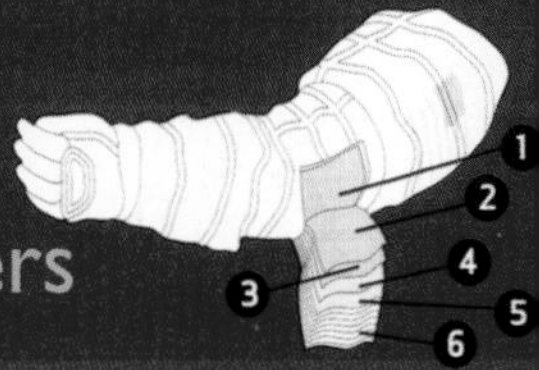

1 Thermal Micrometeoroid Garment (TMG). Cover: Ortho/KEVLAR® reinforced with GORE-TEX®.
2 TMG Insulation. Five to seven layers of aluminized Mylar® (more layers on arms and legs).
3 TMG liner. Neoprene-coated nylon ripstop.
4 Pressure garment cover. Restraint: Dacron®.
5 Pressure garment bladder. Urethane-coated nylon oxford fabric.
6 Liquid cooling garment. Neoprene tubing.

Suit's nominal pressure	0.3 atm (4.3 psi)
Atmosphere	100% oxygen
Primary oxygen tank pressure	900 psi
Secondary oxygen tank pressure	6,000 psi (30-min backup supply)
Maximum EVA duration	8 h
Mass of entire EMU	178 kg (393 lb)
Suit life	30 yr

The Simplified Aid For EVA Rescue (SAFER) provides a compressed nitrogen-powered backpack that permits a crewmember to maneuver independently of the ISS. Its principal use is that it allows a crewmember to maneuver back to the Station if he or she becomes detached from the ISS.

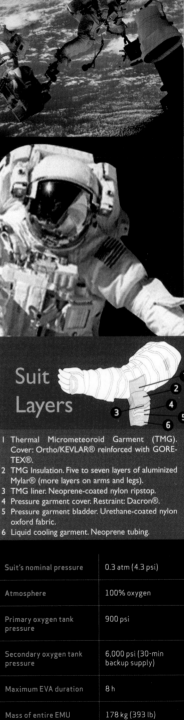

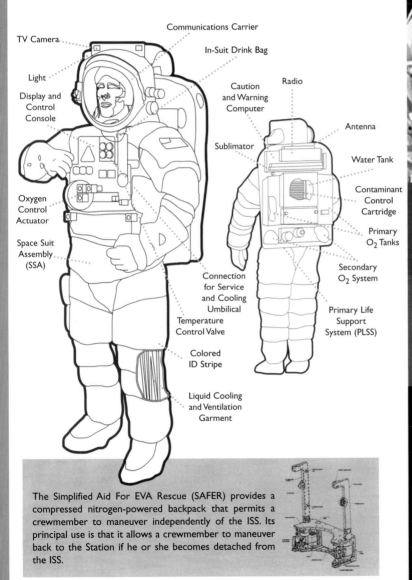

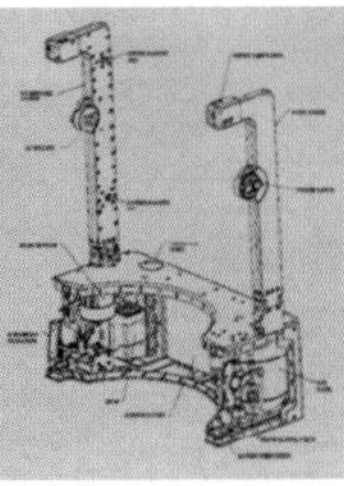

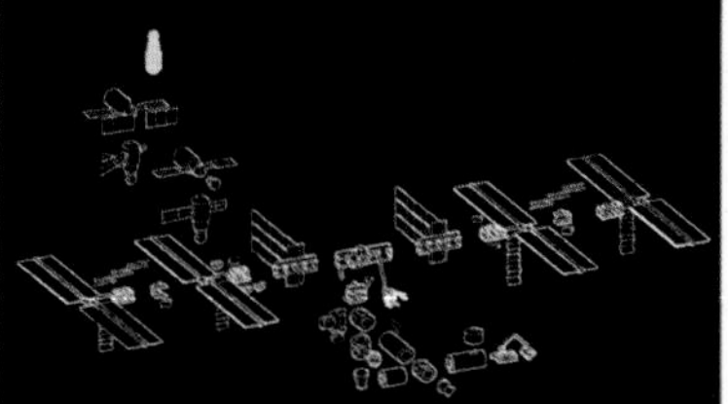

Russian Docking Compartment (DC) and Airlock (Pirs [Pier])

S.P. Korolev Rocket and Space Corporation Energia (RSC Energia)

Pirs provides the capability for extravehicular activity using Russian Orlan suits. Pirs also provides contingency capability for ingress for U.S. EMU EVAs. Additionally, Pirs provides systems for servicing and refurbishing the Orlan suits. The nadir Docking System on Pirs provides a port for the docking of Soyuz and Progress logistics vehicles. When the final Russian science module arrives, Pirs will be moved to the zenith Service Module port.

View of the zenith end of the DC, with probe extended, as it prepares to dock with the ISS in 2001.

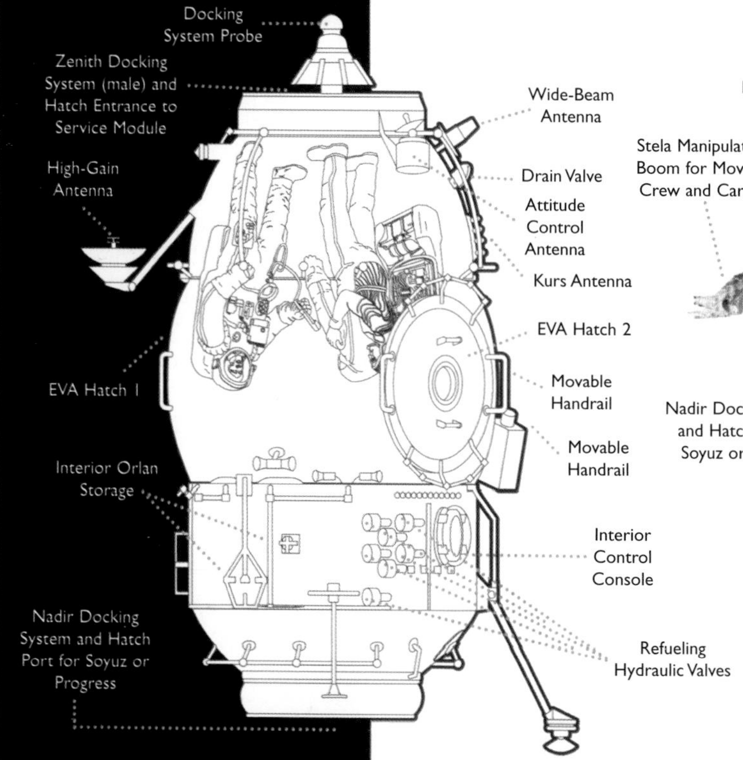

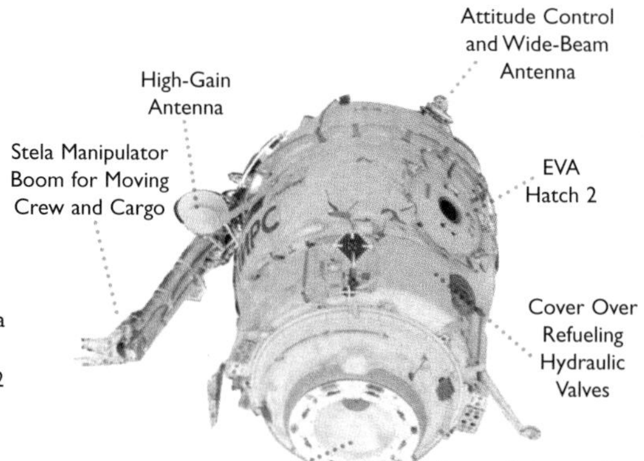

View of the nadir end of the DC.

DC in preparation for launch.

Length	4.9 m (16 ft)
Maximum diameter	2.55 m (8.4 ft)
Mass	3,838 kg (8,461 lb)
Volume	13 m^3 (459 ft^3)
Launch date	August 14, 2001, on Progress M, ISS mission 4R

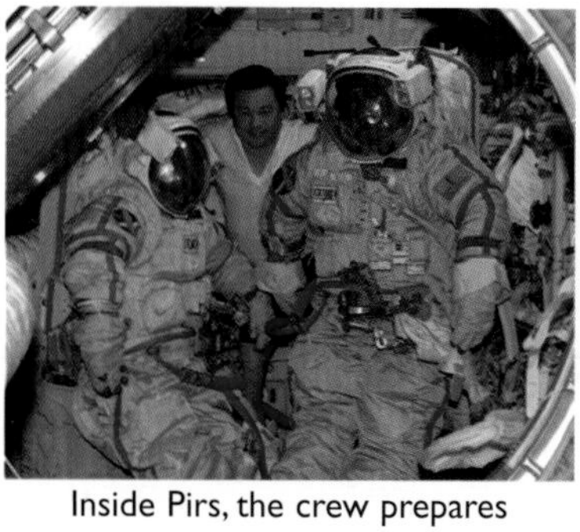

Inside Pirs, the crew prepares Orlan suits for EVA.

Pirs Module location at Service Module nadir.

Orlan Spacesuit

Science Production Enterprise Zvezda

The Orlan-M spacesuit is designed to protect an EVA crewmember from the vacuum of space, ionizing radiation, solar energy, and micrometeoroids. The main body and helmet of the suit are integrated and are constructed of aluminum alloy. Arms and legs are made of a flexible fabric material. Crewmembers enter from the rear via the backpack door, which allows rapid entry and exit without assistance. The Orlan-M spacesuit is a "one-size-fits-most" suit.

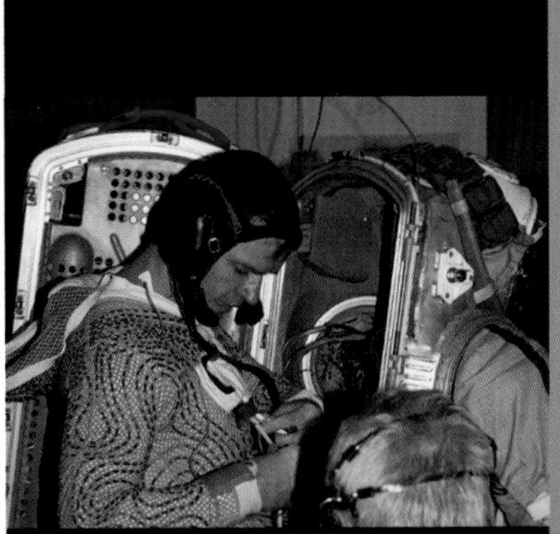

Crewmember in liquid cooling garment prepares to enter Orlan hatch.

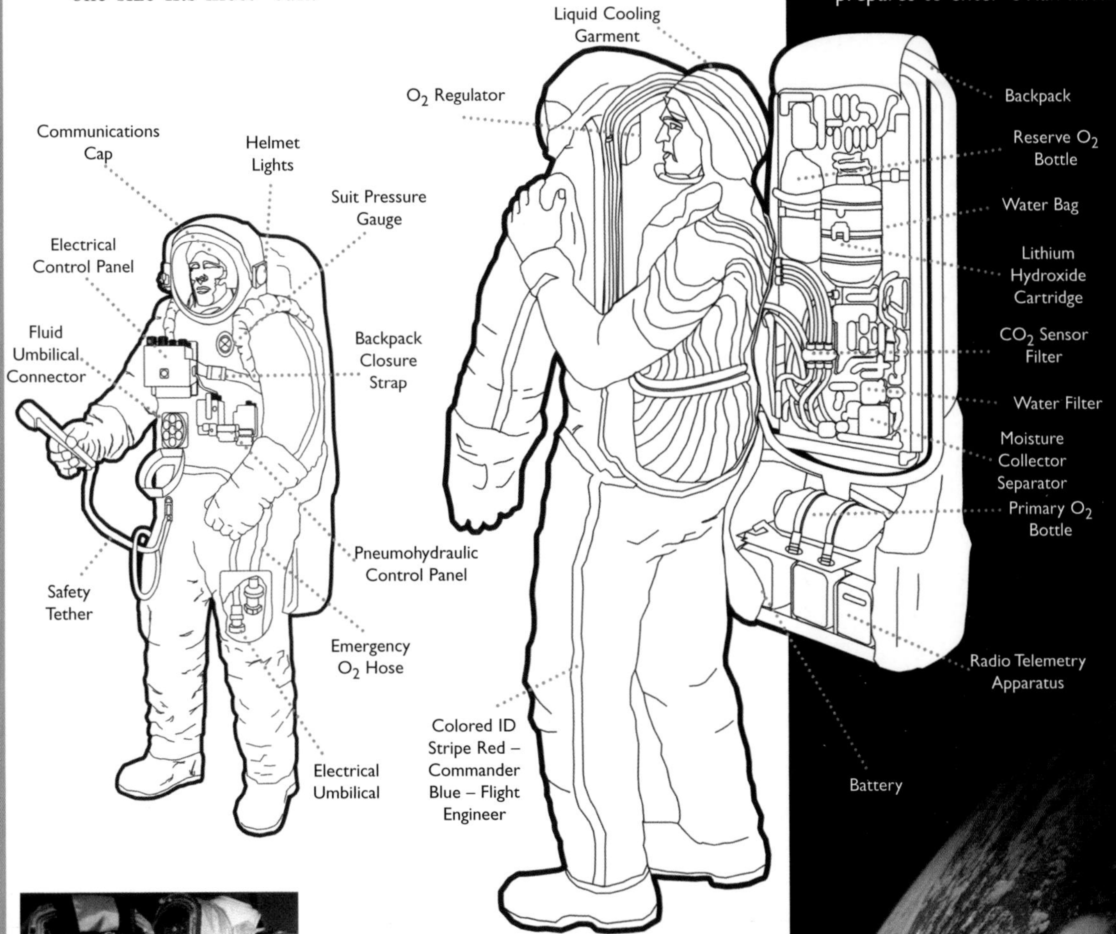

Interior of Orlan suit with rear access hatch open.

The suit operates at a nominal 0.4 atm (5.8 psi) with a 100% oxygen atmosphere.

The suit's maximum EVA duration is 7 hours.

The weight of the entire Orlan assembly is 238 lb.

Orlan is designed for an on-orbit lifetime of 12 EVAs or 4 years without return to Earth.

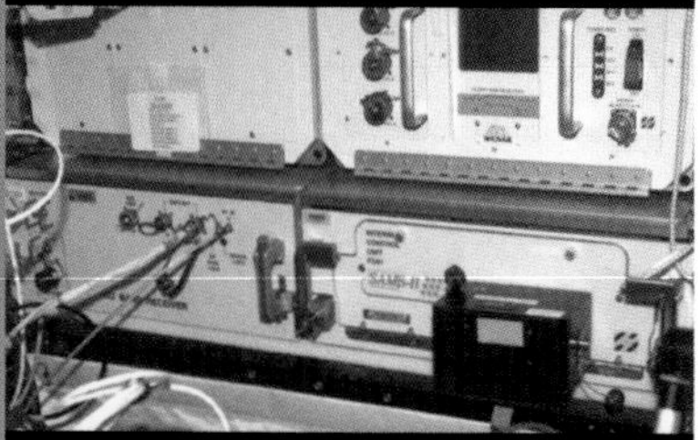

Ku band radio in U.S. Lab.

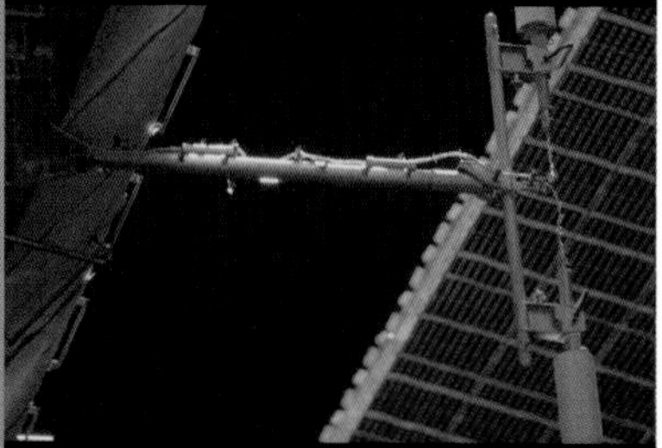

UHF antenna on the P1 Truss.

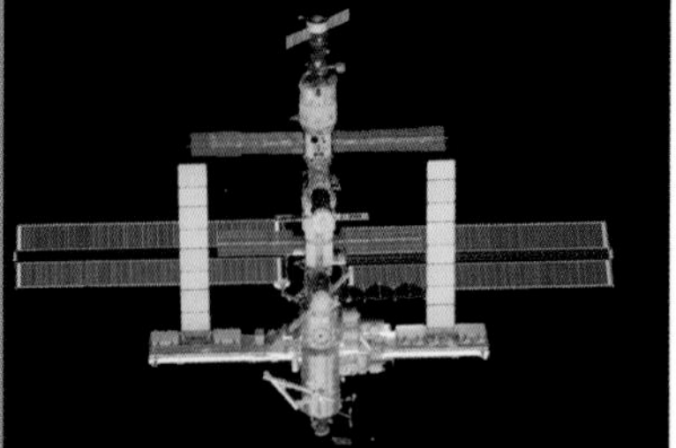

ISS configuration, 2003–2006.

Yuri Onofrienko during communications pass.

Tammy Jernigan wearing EMU communications carrier ("Snoopy cap").

Communications

The radio and satellite communications network allows ISS crews to talk to the ground control centers and the orbiter. It also enables ground control to monitor and maintain ISS systems and operate payloads, and it permits flight controllers to send commands to those systems. The network routes payload data to the different control centers around the world.

The communications system provides the following:

- Two-way audio and video communication among crewmembers aboard the ISS, including crewmembers who participate in an extravehicular activity (EVA);
- Two-way audio, video, and file transfer communication between the ISS and flight control teams located in the Mission Control Center-Houston (MCC-H), other ground control centers, and payload scientists on the ground;
- Transmission of system and payload telemetry from the ISS to the MCC-H and the Payload Operations Center (POC);
- Distribution of ISS experiment data through the POC to payload scientists; and
- Control of the ISS by flight controllers through commands sent via the MCC-H.

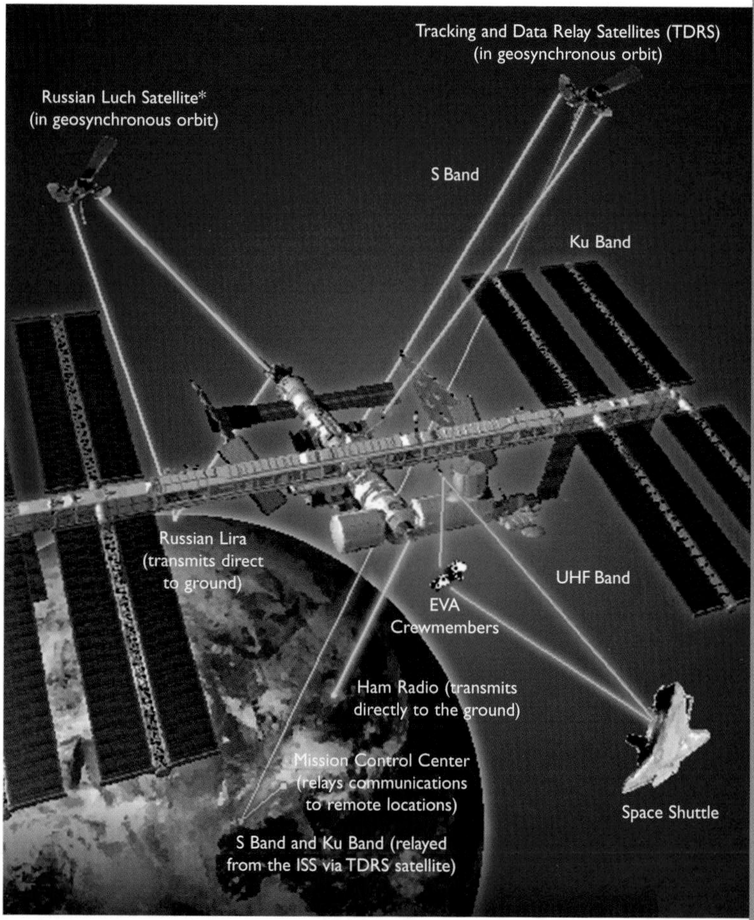

* Luch not currently in use.

Guidance, Navigation, and Control (GN&C)

The International Space Station is a large, free-flying vehicle. The attitude or orientation of the ISS with respect to Earth and the Sun must be controlled; this is important for maintaining thermal, power, and microgravity levels, as well as for communications.

The GN&C system tracks the Sun, communications and navigation satellites, and ground stations. Solar arrays, thermal radiators, and communications antennas aboard the ISS are pointed using the tracking information.

The preferred method of attitude control is the use of gyrodynes, Control Moment Gyroscopes (CMGs) mounted on the Z1 Truss segment. CMGs are 98-kilogram (220-pound) steel wheels that spin at 6,600 revolutions per minute (rpm). The highrotation velocity and large mass allow a considerable amount of angular momentum to be stored. Each CMG has gimbals and can be repositioned to any attitude. As the CMG is repositioned, the resulting force causes the ISS to move. Using multiple CMGs permits the ISS to be moved to new positions or permits the attitude to be held constant. The advantages of this system are that it relies on electrical power generated by the solar arrays and that it provides smooth, continuously variable attitude control. CMGs are, however, limited in the amount of angular momentum they can provide and the rate at which they can move the Station. When CMGs can no longer provide the requisite energy, rocket engines are called upon.

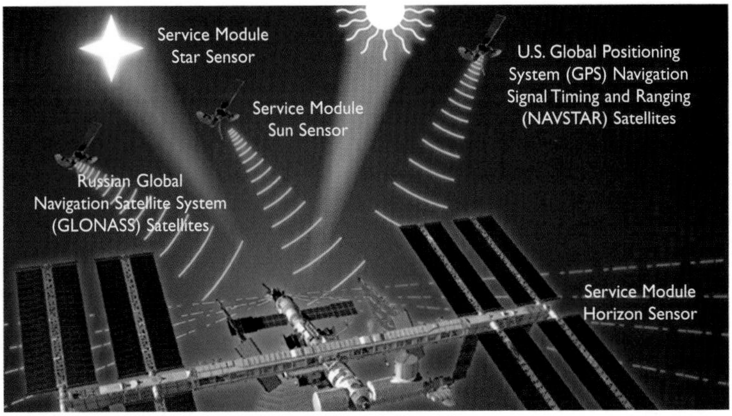

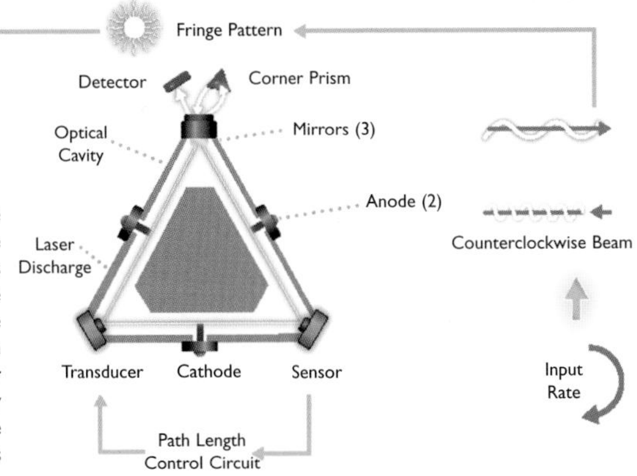

The Rate Gyroscope Assemblies (RGAs) are the U.S. attitude rate sensors used to measure the changing orientation of the ISS. RGAs are installed on the back of the Truss, under the GPS antennas, and they are impossible to see on the ISS unless shielding is

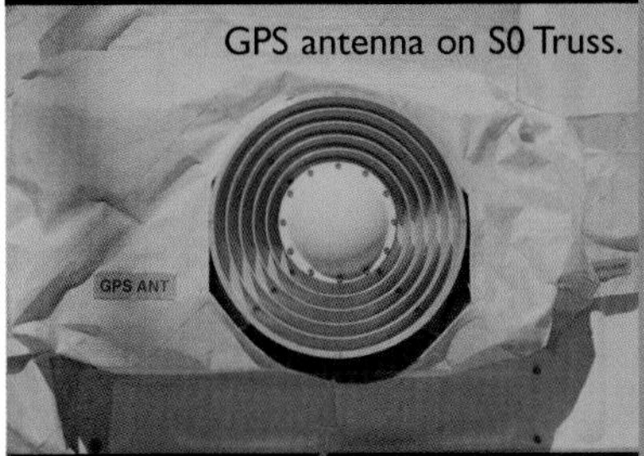
GPS antenna on S0 Truss.

Control Moment Gyroscopes on the Z1 Truss.

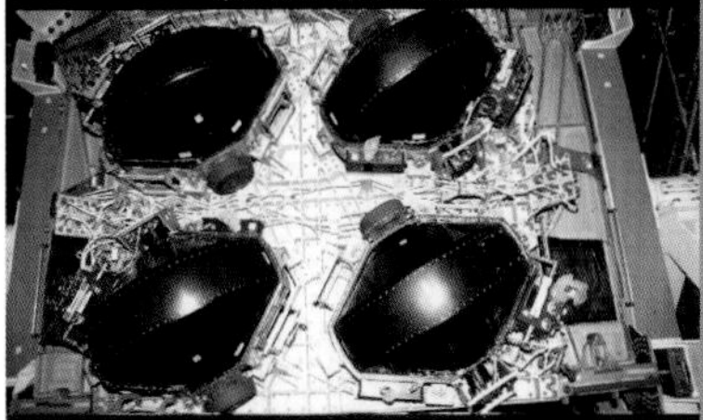
Control Moment Gyroscope gimbals used for orienting the ISS.

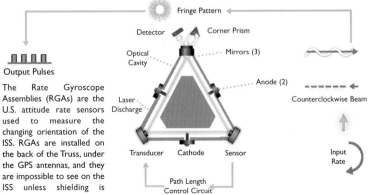

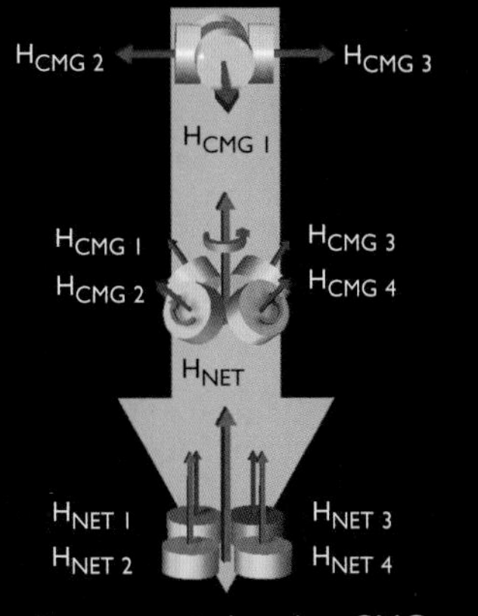

Forces are induced as CMGs are repositioned.

Electrical Power Systems (EPS)

The EPS generates, stores, and distributes power and converts and distributes secondary power to users.

Crewmember Mike Fincke replaces the Remote Power Controller Module (RPCM) on the S0 Truss.

Solar Array Wing (SAW) (has 2 arrays and 32,800 solar cells; converts sunlight to DC power, producing a maximum of 31 kW at the beginning of its life and degrading to 26 kW after 15 years; each cell is approximately 14% efficient, which was state-of-the-art at the time of design)

Photovoltaic Radiator (circulates cooling fluid to maintain EPS/battery temperature)

Nickel-Hydrogen Batteries (store electrical energy for use during the night; Battery Charge Discharge Unit [BCDU] controls each battery's charge)

Solar (Array) Alpha Rotation Joint (SARJ) (tracks the Sun throughout Earth orbit)

Sunlight

Main Bus Switching Units (MBSUs) (route power to proper locations in the ISS)

Direct Current (DC) Switching Unit (DCSU) (routes power from the solar array to the MBSUs in the S0 Truss that control power to different ISS locations)

Sequential Shunt Unit (SSU) (maintains constant voltage at 160 V, sending excess power back to array)

Beta Gimbal (used for tracking the Sun because of seasonal changes)

MBSUs

Remote Power Controllers (RPCs) (control the flow of electric power to users)

Integrated Equipment Assembly (IEA) Truss (houses EPS hardware)

Electronic Control Unit (ECU) (controls pointing of solar arrays)

U.S. Lab

DC-to-DC Converter Units (DDCUs) – Some Located on Truss (convert primary 160-V power to secondary 124-V power)

Power going from MBSUs through umbilicals into U.S. Lab

DDCUs – Some located in modules (convert primary 160-V power to secondary 124-V power)

Crewmember Mike Fincke holds an RPCM in the Quest Airlock. It was later used to replace an RPCM on the S0 Truss.

Solar Array

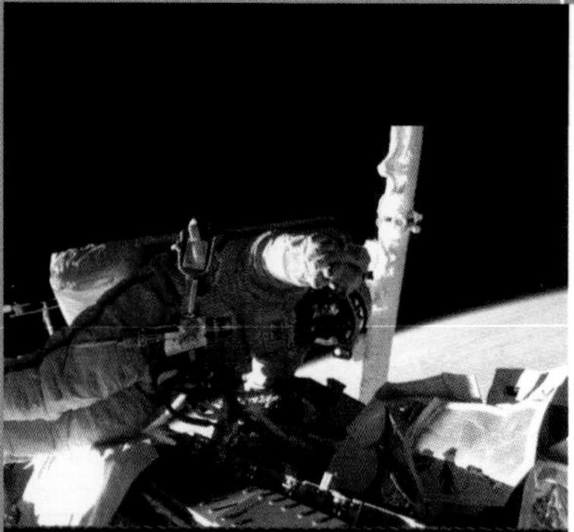

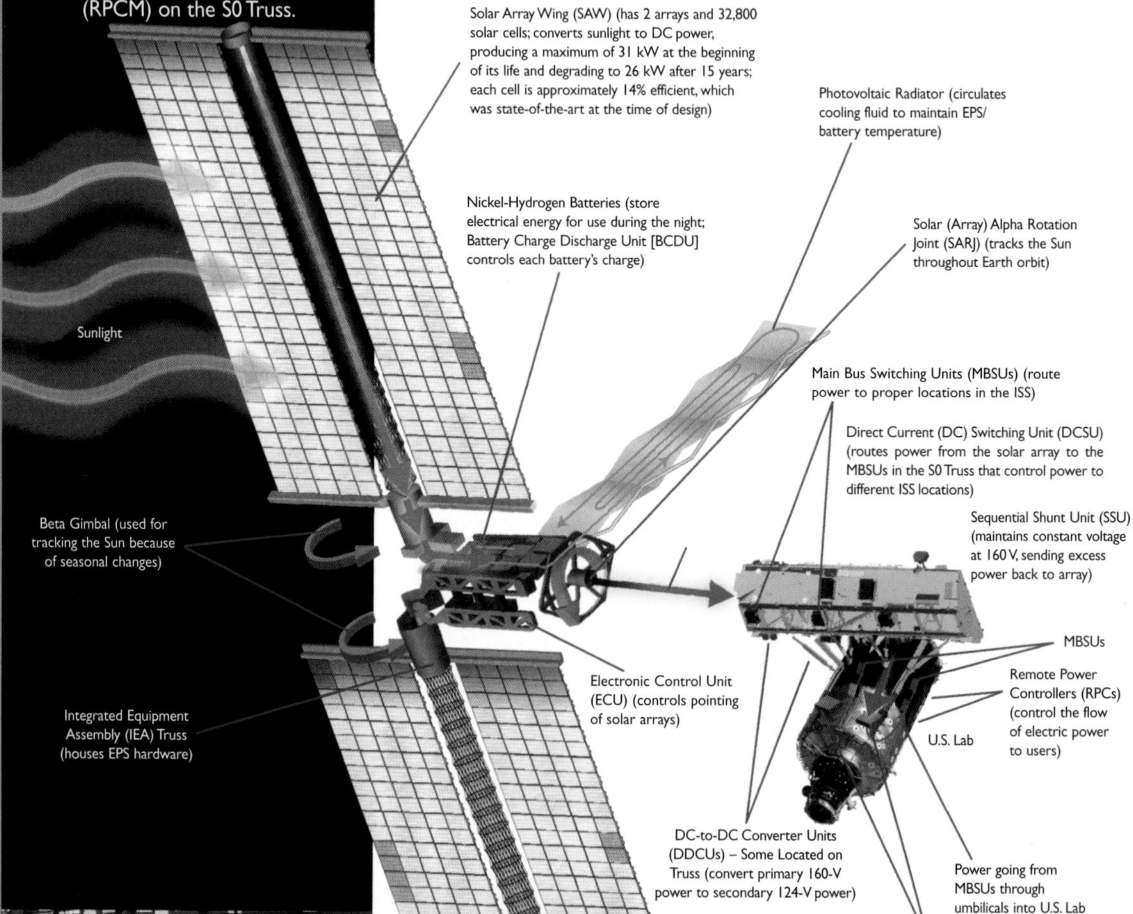

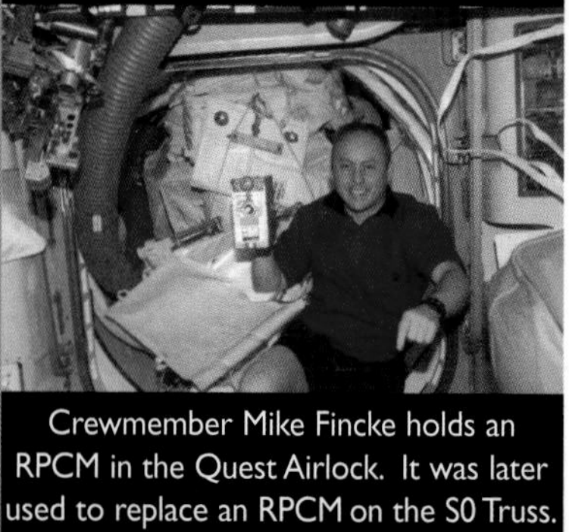

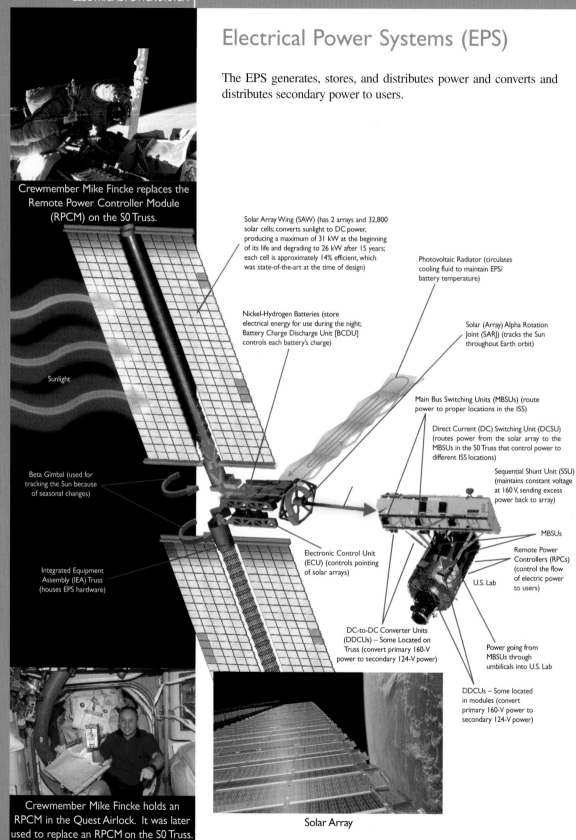

Communications

The radio and satellite communications network allows ISS crews to talk to the ground control centers and the orbiter. It also enables ground control to monitor and maintain ISS systems and operate payloads, and it permits flight controllers to send commands to those systems. The network routes payload data to the different control centers around the world.

The communications system provides the following:

- Two-way audio and video communication among crewmembers aboard the ISS, including crewmembers who participate in an extravehicular activity (EVA);
- Two-way audio, video, and file transfer communication between the ISS and flight control teams located in the Mission Control Center-Houston (MCC-H), other ground control centers, and payload scientists on the ground;
- Transmission of system and payload telemetry from the ISS to the MCC-H and the Payload Operations Center (POC);
- Distribution of ISS experiment data through the POC to payload scientists; and
- Control of the ISS by flight controllers through commands sent via the MCC-H.

Port and Starboard Radiator panels from truss below U.S. solar array.

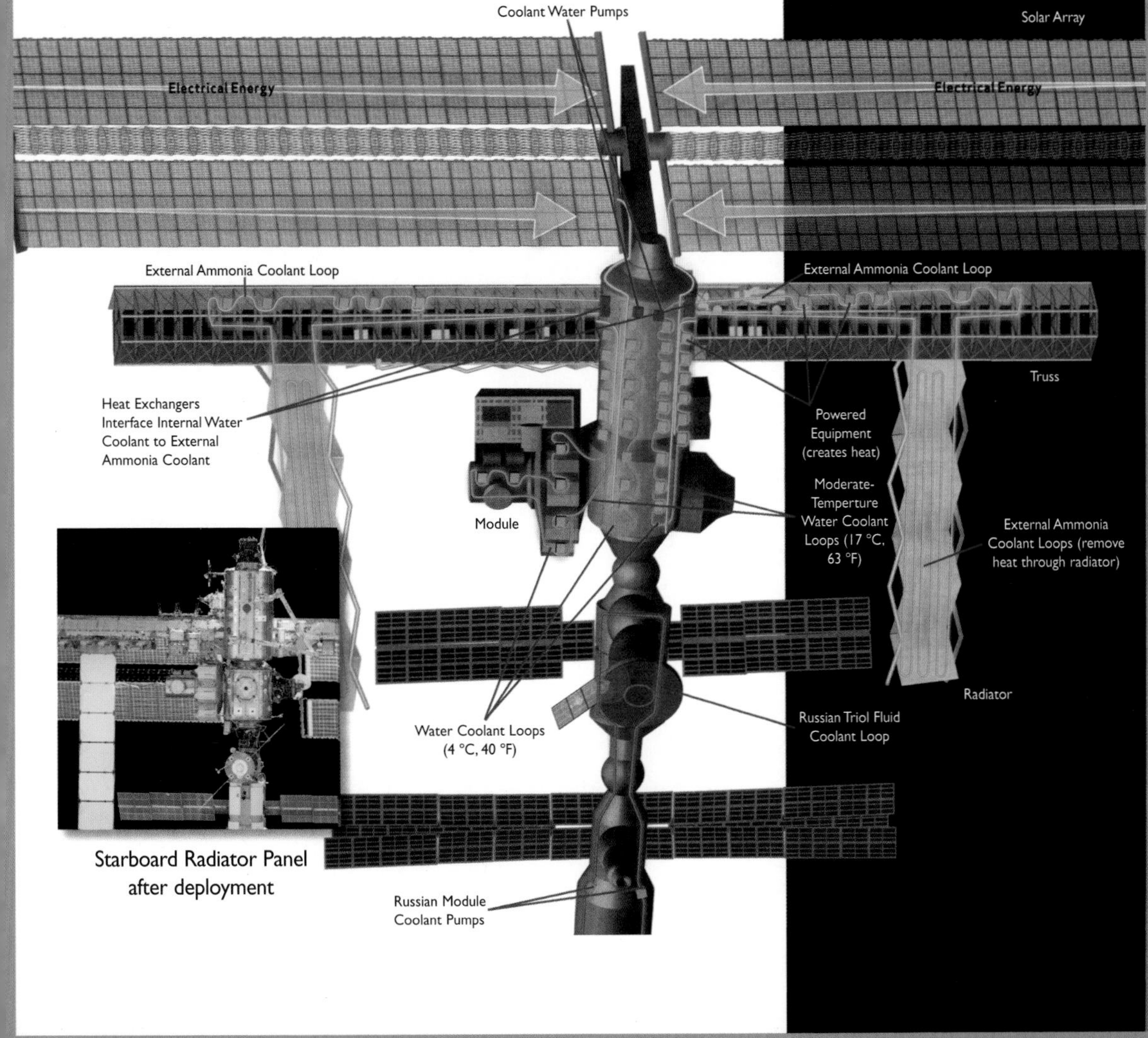

Starboard Radiator Panel after deployment

NASA
esa
P K
CIO
OPS
TOR
NASA
IBM

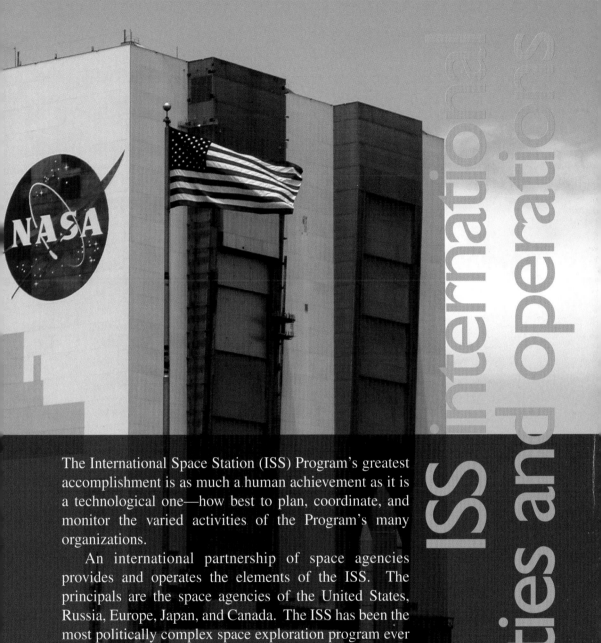

ISS international facilities and operations

The International Space Station (ISS) Program's greatest accomplishment is as much a human achievement as it is a technological one—how best to plan, coordinate, and monitor the varied activities of the Program's many organizations.

An international partnership of space agencies provides and operates the elements of the ISS. The principals are the space agencies of the United States, Russia, Europe, Japan, and Canada. The ISS has been the most politically complex space exploration program ever undertaken.

(continued on the next page)

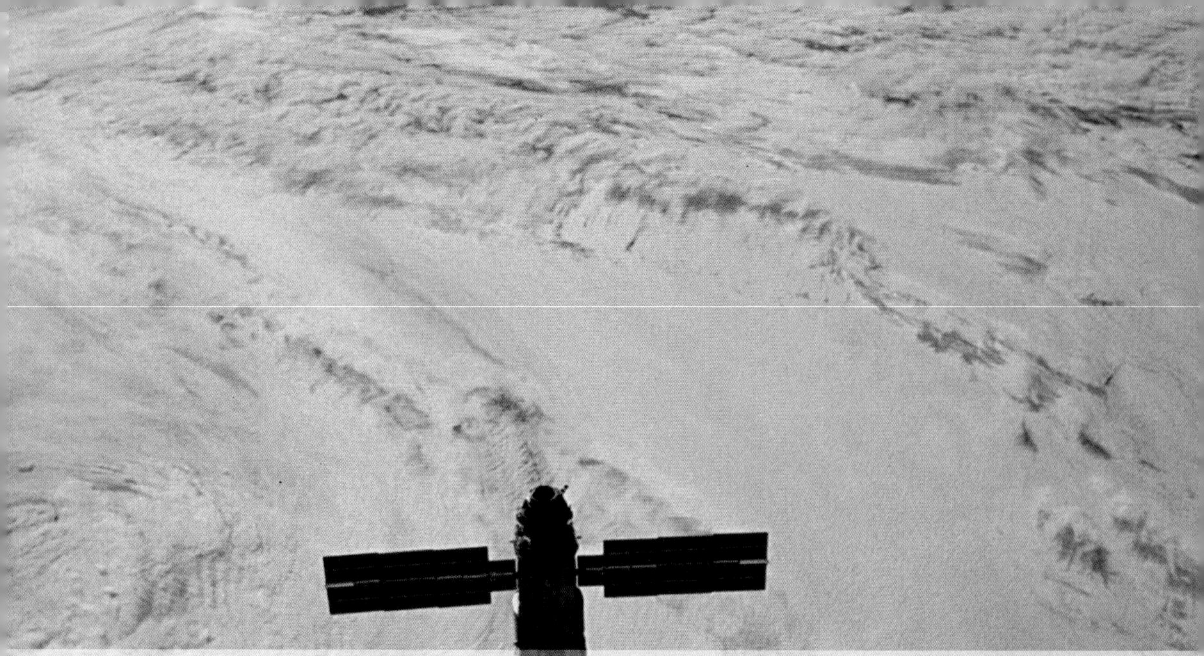

(continued from the previous page)

The ISS Program brings together international flight crews; multiple launch vehicles; globally distributed launch, operations, training, engineering, and development facilities; communications networks; and the international scientific research community. Elements launched from different countries and continents are not mated together until they reach orbit, and some elements that have been launched later in the assembly sequence were not yet built when the first elements were placed in orbit.

Operating the ISS is even more complicated than other space flight endeavors because it is an international program. Each ISS partner has the primary responsibility to manage and run the hardware it provides. But the various elements provided by the ISS partners are not independent, and over time they must be operated as an integrated system.

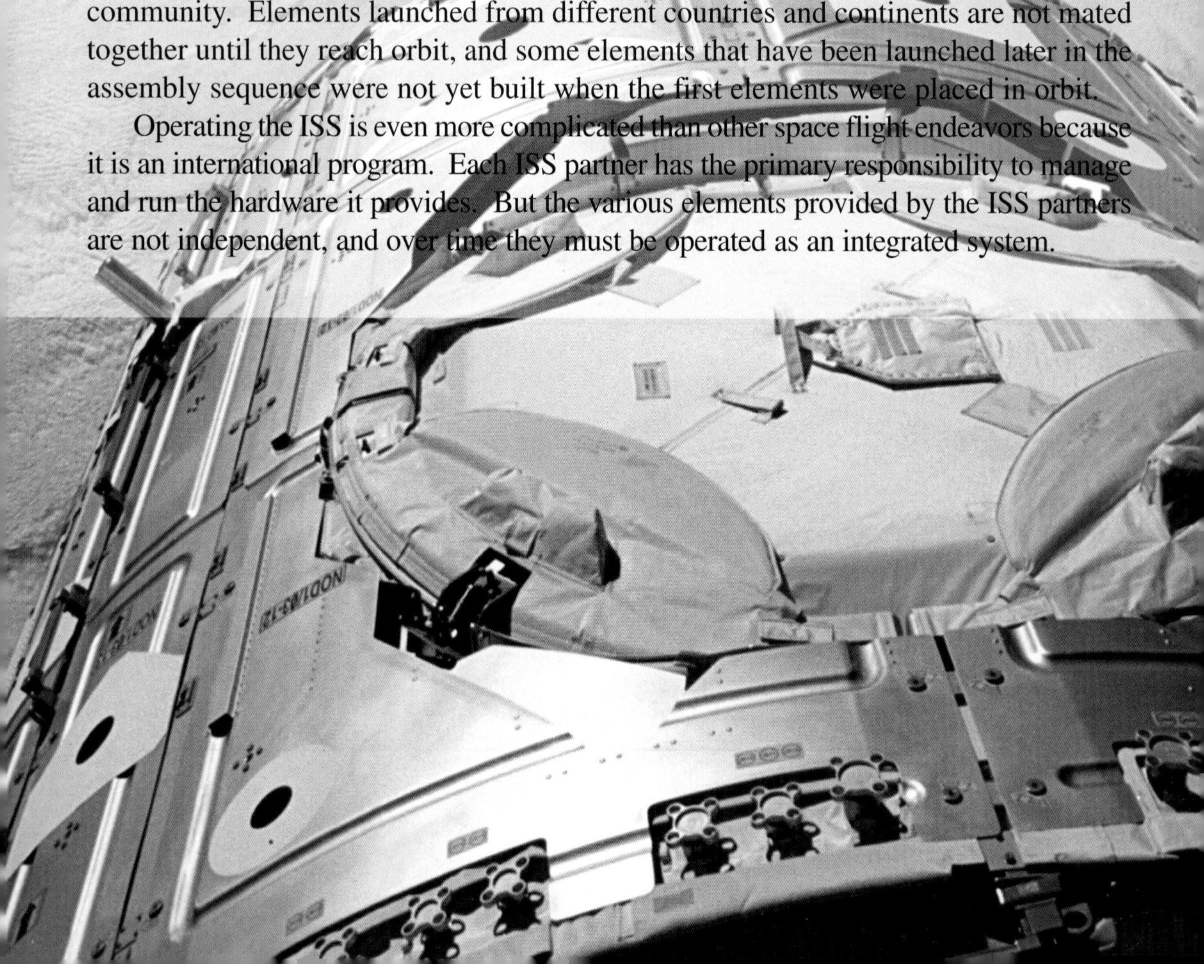

National Aeronautics and Space Administration
United States

Canadian Space Agency

European Space Agency

Japan Aerospace Exploration Agency

Russian Federal Space Agency

ISS Operations and Management

Glenn Telescience Support Center
Cleveland, Ohio, U.S.
CSA Headquarters, Mobile Servicing System (MSS) Control and Training
Saint-Hubert, Quebec, Canada
NASA Headquarters
Washington, DC, U.S.
Payload Operations Center (POC)
Huntsville, Alabama, U.S.
Ames Telescience Support Center
Moffett Field, California, U.S.
Shuttle Launch Control
Kennedy Space Center, Florida, U.S.
ISS Training
ISS Program Management
ISS Mission Control
Houston, Texas, U.S.
Ariane Launch Control
Kourou, French Guiana

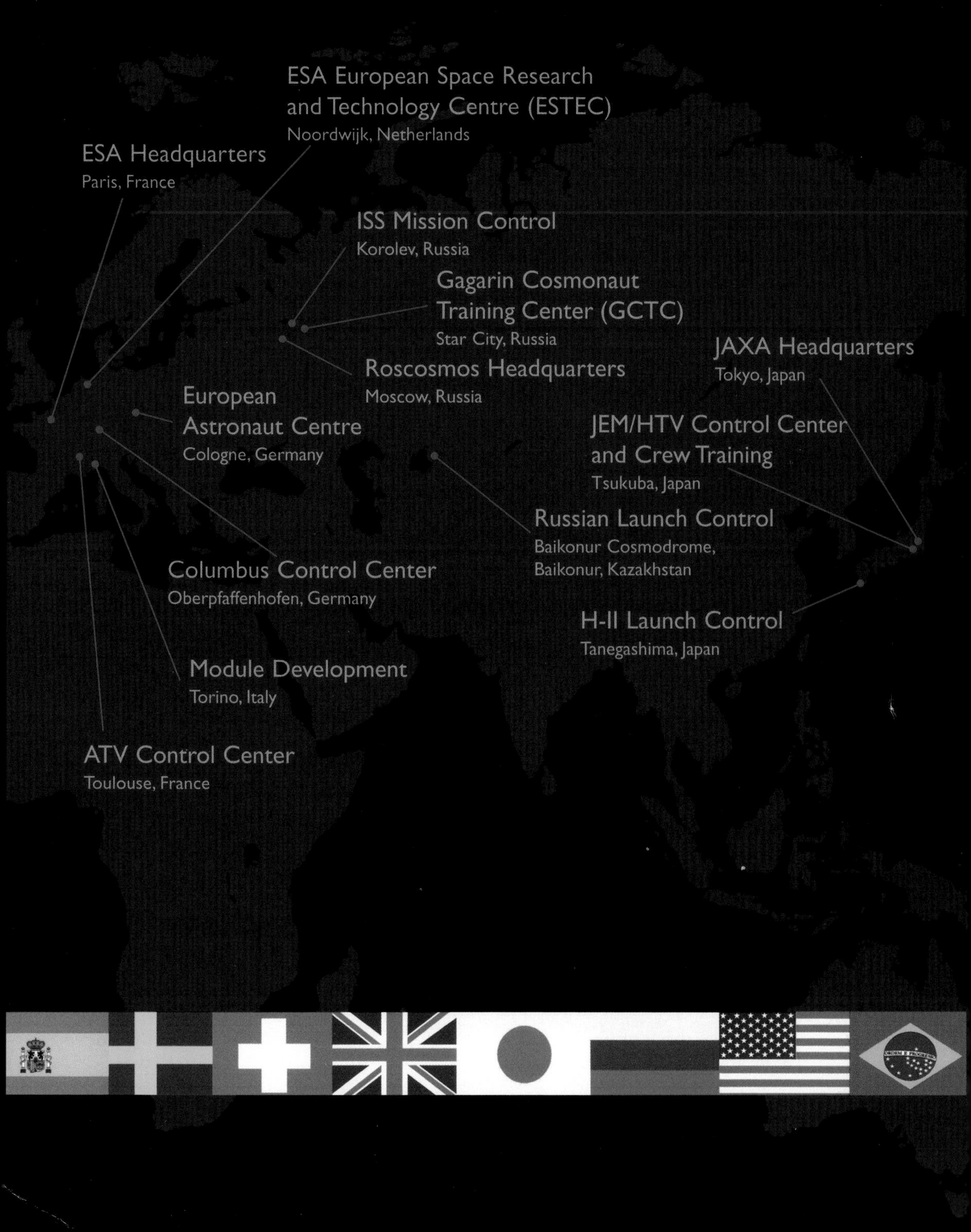
ESA European Space Research and Technology Centre (ESTEC)
Noordwijk, Netherlands
ESA Headquarters
Paris, France
ISS Mission Control
Korolev, Russia
Gagarin Cosmonaut Training Center (GCTC)
Star City, Russia
Roscosmos Headquarters
Moscow, Russia
JAXA Headquarters
Tokyo, Japan
European Astronaut Centre
Cologne, Germany
JEM/HTV Control Center and Crew Training
Tsukuba, Japan
Russian Launch Control
Baikonur Cosmodrome, Baikonur, Kazakhstan
Columbus Control Center
Oberpfaffenhofen, Germany
H-II Launch Control
Tanegashima, Japan
Module Development
Torino, Italy
ATV Control Center
Toulouse, France

United States of America

National Aeronautics and Space Administration (NASA)

NASA HEADQUARTERS (HQ)

NASA headquarters, in Washington, DC, exercises management over the NASA field centers, establishes management policies, and analyzes all phases of the ISS Program.

JOHNSON SPACE CENTER (JSC)

Johnson Space Center, in Texas, directs the ISS Program.Mission Control manages activities aboard the U.S. segment of the ISS. JSC is the primary Center for spacecraft design, development, and mission integration. JSC is also the primary location for crew training.

KENNEDY SPACE CENTER (KSC)

Kennedy Space Center, in Florida, prepares the ISS modules and Space Shuttle orbiters for each mission, coordinates each countdown, and manages Space Shuttle launch and post-landing operations.

MARSHAll SPACE FLIGHT CENTER (MSFC)

Marshall Space Flight Center's Payload Operation Center (POC) is the ground control center for experiments and payloads being operated on the ISS. MSFC has also overseen development of most U.S. modules and the ISS EClSS system.

TELESCIENCE SUPPORT CENTERS (TSCs)

Telescience Support Centers around the country are equipped to conduct science operations on board the ISS. These TSCs are located at Marshall Space Flight Center in huntsville, Alabama; Ames Research Center (ARC) in Moffett Field, California; Glenn Research Center (GRC) in Cleveland, Ohio; and Johnson Space Center in Houston, Texas.

DESIGN, DEVELOPMENT, TESTING, EVALUATION, and INTEGRATION (DDTE&I)

Boeing is NASA's prime ISS contractor. It oversees the development, testing, and preparation for launch of the ISS elements.

http://www.nasa.gov

Canada

Canadian Space Agency (CSA)

MOBILE SERVICING SYSTEM (MSS) OPERATIONS COMPLEX (MOC)

The MSS Operations Complex in Longueuil, Quebec, provides the resources, equipment, and expertise needed for the engineering and monitoring of the Mobile Servicing System as well as for crew training.

SPACE STATION REMOTE MANIPULATOR SYSTEM (SSRMS) DESIGN AND DEVELOPMENT

The SSRMS was designed and developed by Macdonald, Dettwiler and Associates, Ltd., in Brampton, Ontario.

http://www.space.gc.ca

Europe

European Space Agency (ESA)

EUROPEAN SPACE RESEARCH AND TECHNOLOGY CENTRE (ESTEC)

The European Space Research and Technology Centre, the largest site and the technical heart of the ESA, is in Noordwijk, in the Netherlands. Most ESA projects are developed here by more than 2,000 specialists.

COLUMBUS CONTROL CENTRE (COL-CC) AND AUTOMATED TRANSFER VEhICIE CONTROl CENTRE (ATV-CC)

Two ground control centers are responsible for controlling and operating the European contribution to the ISS program. These are the Columbus Control Centre and the Automated Transfer vehicle Control Centre. The COL-CC, located at the German Aerospace Center (DLR), in Oberpfaffenhofen, near Munich, Germany, will control and operate the Columbus Research Laboratory and coordinate European experiments (payload) operations. The ATC-CC, located in Toulouse, France, on the premises of the French space agency Centre national d 'Etudes Spatiales (CnES), will control and operate the ATVs.

GUIANA SPACE CENTRE (GSC)

Europe's Spaceport is situated in the northeast of South America in French Guiana. Initially created by CNES, it is jointly funded and used by both the French space agency and ESA as the launch site for the Ariane 5 vehicle.

EUROPEAN ASTRONAUT CENTRE (EAC)

The European Astronaut Centre of the European Space Agency is situated in Cologne, Germany.I t was established in 1990 and is the home base of the 13 European astronauts who are members of the European Astronaut Corps.

USER CENTERS

User Support and Operation Centers (USOCs) are based in national centers distributed throughout Europe. These centers are responsible for the use and implementation of European payloads aboard the ISS.

http://www.esa.int

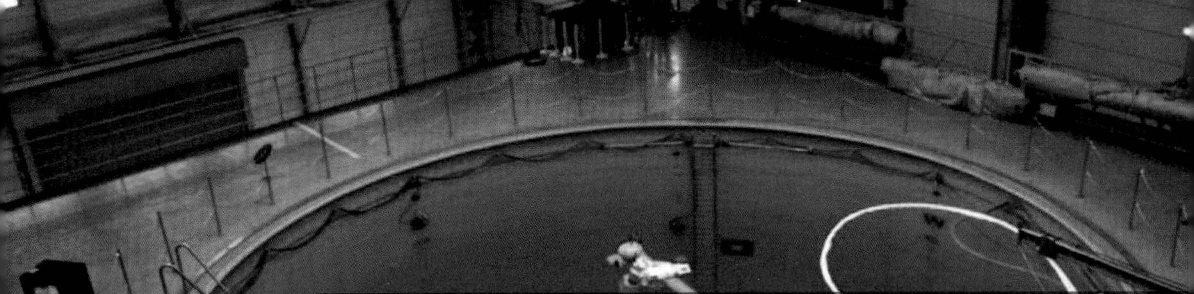

Japan

Japan Aerospace Exploration Agency (JAXA)

In addition to the JAXA headquarters in Tokyo and other field centers throughout the country, Tsukuba Space Center and Tanegashima launch Facility are JAXA's primary ISS facilities.

TSUKUBA SPACE CENTER (TKSC)

JAXA's Tsukuba Space Center is located in Tsukuba Science City. As part of the International Space Station project, the Japanese Experiment Module (JEM) "Kibo" is developed and tested at TKSC. JAXA is preparing the Kibo Control Centre for support of the JEM once it is launched. Astronaut training for JEM will be conducted at JAXA.

TANEGASHIMA SPACE CENTER (TNSC)

The Tanegashima Space Center is the largest space-development facility in Japan and is located in the south of Kagoshima Prefecture, along the southeast coast of Tanegashima. The Osaki Range is onsite for J-I and H-IIA launch vehicles. There are also related developmental facilities for test firings of liquid-and solid-fuel rocket engines.

http://www.jaxa.jp/index_e.html

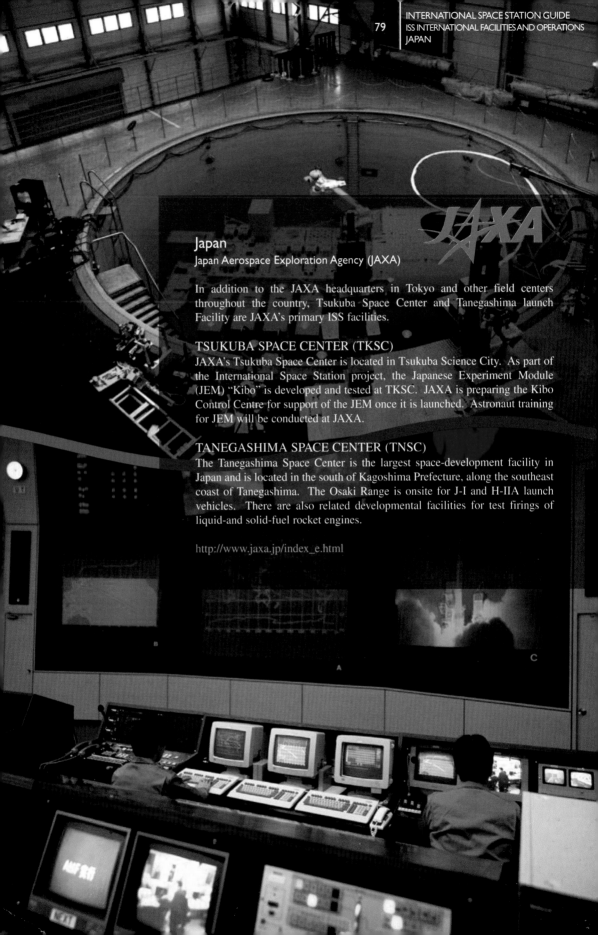

Russia

Roscosmos, the Russian Federal Space Agency

Roscosmos oversees all Russian human space flight activities.

MOSCOW MISSION CONTROL (TSUP)

Moscow Mission Control is the primary Russian facility for the control of human space flight. It is located in Korolev, outside of Moscow.

GAGARIN COSMONAUT TRAINING CENTER (GCTC)

The gagarin Cosmonaut Training Center, at Zvezdny gorodok (Star City), provides full-size trainers and simulators of all Russian ISS modules, a water pool used for spacewalk training, centrifuges to simulate g-forces during liftoff, and a planetarium used for celestial navigation.

S.P. KOROLEV ROCKET AND SPACE CORPORATION ENERGIA (RSC ENERGIA)

RSC Energia, in Korolev, outside of Moscow, integrates spacecraft hardware and manages the ISS Program implementation for the Russian segment.

KHRUNICHEV STATE RESEARCH AND PRODUCTION SPACE CENTER (KHRUNICHEV)

Khrunichev, in Moscow, is the prime contractor for the Functional Cargo Block, Service Module, and Proton launch vehicles.

SCIENCE PRODUCTION ENTERPRISE ZVEZDA

Science Production Enterprise Zvezda, in Tomolino, near Moscow, is the primary developer of the Russian Orlan and Sokol spacesuits that are used for the ISS.

BAIKONUR COSMODROME

The Baikonur Cosmodrome, in Kazakhstan, is the chief launch center for both piloted and unpiloted space vehicles. It supports the Soyuz and Proton launch vehicles and plays an essential role in the deployment and operation of the International Space Station.

INSTITUTE FOR BIOMEDICAL PROBLEMS (IBMP)

The Institute for Biomedical Problems, outside Moscow, conducts scientific research and develops hardware for the protection of crew health.

http://www.roscosmos.ru

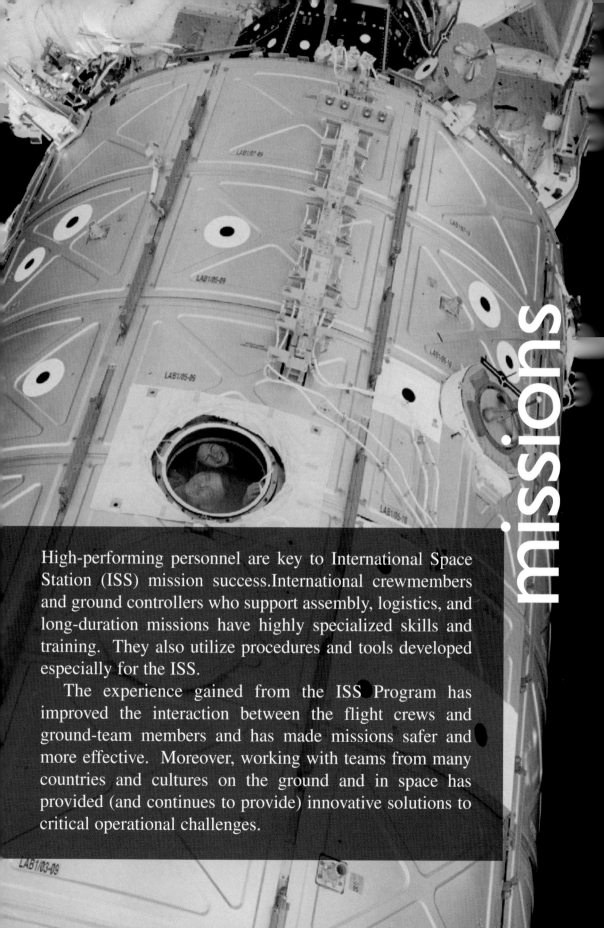

missions

High-performing personnel are key to International Space Station (ISS) mission success.International crewmembers and ground controllers who support assembly, logistics, and long-duration missions have highly specialized skills and training. They also utilize procedures and tools developed especially for the ISS.

The experience gained from the ISS Program has improved the interaction between the flight crews and ground-team members and has made missions safer and more effective. Moreover, working with teams from many countries and cultures on the ground and in space has provided (and continues to provide) innovative solutions to critical operational challenges.

ISS Expeditions and Crews

Expedition	Patch	Crew
Expedition 1		
Expedition 2		
Expedition 3		
Expedition 4		
Expedition 5		
Expedition 6		
Expedition 7		
Expedition 8		
Expedition 9		
Expedition 10		
Expedition 11		
Expedition 12		
Expedition 13		
Expedition 14		

Expedition 1
William Shepherd, U.S.
Yuri Gidzenko, Russia
Sergei Krikalev, Russia
Launched: Oct. 2000
Returned: Mar. 2001
136 days on ISS (141 in space)

Expedition 2
Yuri Usachev, Russia
Jim Voss, U.S.
Susan Helms, U.S.
Launched: Mar. 2001
Returned: Aug. 2001
163 days on ISS (167 in space)

Expedition 3
Frank Culbertson, U.S.
Vladimir Dezhurov, Russia
Mikhail Tyurin, Russia
Launched: Aug. 2001
Returned: Dec. 2001
125 days on ISS (129 in space)

Expedition 4
Yury Onufrienko, Russia
Carl Walz, U.S.
Daniel Bursch, U.S.
Launched: Dec. 2001
Returned: June 2002
190 days on ISS (196 in space)

Expedition 5
Valery Korzun, Russia
Sergei Treschev, Russia
Peggy Whitson, U.S.
Launched: June 2002
Returned: Dec. 2002
178 days on ISS (185 in space)

Expedition 6
Kenneth Bowersox, U.S.
Nikolai Budarin, Russia
Donald Pettit, U.S.
Launched: Nov. 2002
Returned: May 2003
159 days on ISS (161 in space)

Expedition 7
Yuri Malenchenko, Russia
Edward Lu, U.S.
Launched: Apr. 2003
Returned: Oct. 2003
183 days on ISS (185 in space)

Expedition 8
Michael Foale, U.S.
Alexander Kaleri, Russia
Launched: Oct. 2003
Returned: Apr. 2004
193 days on ISS (195 in space)

Expedition 9
Gennady Padalka, Russia
E. Michael Fincke, U.S.
Launched: Apr. 2004
Returned: Oct. 2004
186 days on ISS (188 in space)

Expedition 10
Leroy Chiao, U.S.
Salizhan Sharipov, Russia
Launched: Oct. 2004
Returned: Apr. 2005
191 days on ISS (193 in space)

Expedition 11
Sergei Krikalev, Russia
John Phillips, U.S.
Launched: Apr. 2005
Returned: Oct. 2005
177 days on ISS (179 in space)

Expedition 12
William McArthur, U.S.
Valery Tokarev, Russia
Launched: Sept. 2005
Returned: Apr. 2006
184 days on ISS (190 in space)

Expedition 13
Pavel Vinogradov, Russia
Jeffrey Williams, U.S.
Thomas Reiter, Germany (start July 2006)
Launched: Apr. 2006
Returned: Sept. 2006 (projected)

Expedition 14
Michael Lopez-Alegria, U.S.
Mikhail Tyurin, Russia
Sunita Williams, U.S.
Scheduled: Sept. 2006–Mar. 2007

Note: Only professional astronauts participating in ISS functions are included. See page 83 for Space Flight Participants who have visited the ISS.

STS Missions and Crews
Space Shuttle Missions to the International Space Station

Mission | Patch | Crew

STS-88

STS-96

STS-101

STS-106

STS-92

STS-97

STS-88 Endeavour
Robert Cabana, U.S.
Nancy Currie, U.S.
Sergei Krikalev, Russia (Roscosmos)
James Newman, U.S.
Jerry Ross, U.S.
Frederick Sturckow, U.S.
Launched: Dec. 4, 1998
Returned: Dec. 15, 1998

STS-96 Discovery
Kent Rominger, U.S.
Daniel Barry, U.S.
Rick Husband, U.S.
Tamara Jernigan, U.S.
Ellen Ochoa, U.S.
Julie Payette, Canada (CSA)
Valery Tokarev, Russia (Roscosmos)
Launched: May 27, 1999
Returned: June 6, 1999

STS-101 Atlantis
James Halsell, U.S.
Susan Helms, U.S.
Scott Horowitz, U.S.
Yury Usachev, Russia (Roscosmos)
James Voss, U.S.
Mary Weber, U.S.
Jeffrey Williams, U.S.
Launched: May 19, 2000
Returned: May 29, 2000

STS-106 Atlantis
Terrence Wilcutt, U.S.
Scott Altman, U.S.
Daniel Burbank, U.S.
Edward Lu, U.S.
Yuri Malenchenko, Russia (Roscosmos)
Richard Mastracchio, U.S.
Boris Morukov, Russia (Roscosmos)
Launched: Sept. 8, 2000
Returned: Sept. 19, 2000

STS-92 Discovery
Leroy Chiao, U.S.
Brian Duffy, U.S.
Michael Lopez-Alegria, U.S.
William McArthur, U.S.
Pamela Melroy, U.S.
Koichi Wakata, Japan (NASDA)
Peter Wisoff, U.S.
Launched: Oct. 11, 2000
Returned: Oct. 24, 2000

STS-97 Endeavour
Michael Bloomfield, U.S.
Marc Garneau, Canada (CSA)
Brent Jett, U.S.
Carlos Noriega, U.S.
Joseph Tanner, U.S.
Launched: Nov. 30, 2000
Returned: Dec. 11, 2000

STS-98 Atlantis
Kenneth Cockrell, U.S.
Robert Curbeam, U.S.
Marsha Ivins, U.S.
Thomas Jones, U.S.
Mark Polansky, U.S.
Launched: Feb. 7, 2001
Returned: Feb. 20, 2001

STS-102 Discovery
James Wetherbee, U.S.
James Kelly, U.S.
Paul Richards, U.S.
Andrew Thomas, U.S.
Yuri Usachev, Russia (Roscosmos), up*
James Voss, U.S., up*
Susan Helms, U.S., up*
William Shepherd, U.S., down*
Yuri Gidzenko, Russia (Roscosmos), down*
Sergei Krikalev, Russia (Roscosmos), down*
Launched: Mar. 8, 2001
Returned: Mar. 21, 2001

STS-100 Endeavour
Jeffrey Ashby, U.S.
Umberto Guidoni, Italy (ESA)
Chris Hadfield, Canada (CSA)
Scott Parazynski, U.S.
John Phillips, U.S.
Kent Rominger, U.S.
Yuri Lonchakov, Russia (Roscosmos)
Launched: Apr. 19, 2001
Returned: May 1, 2001

Mission	Patch	Crew
STS-98		
STS-102		
STS-100		
STS-104		
STS-105		
STS-108		
STS-110		
STS-111		

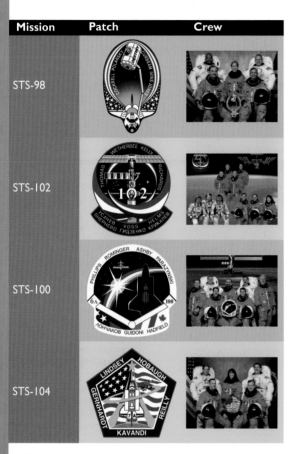

STS-104 Atlantis

Michael Gernhardt, U.S.
Charles Hobaugh, U.S.
Janet Kavandi, U.S.
Steven Lindsey, U.S.
James Reilly, U.S.
Launched: July 12, 2001
Returned: July 24, 2001

STS-105 Discovery

Daniel Barry, U.S.
Patrick Forrester, U.S.
Scott Horowitz, U.S.
Frederick Sturckow, U.S.
Frank Culbertson, U.S., up*
Vladimir Dezhurov, Russia (Roscosmos), up*
Mikhail Turin, Russia (Roscosmos), up*
Yuri Usachev, Russia (Roscosmos), down*
James Voss, U.S., down*
Susan Helms, U.S., down*
Launched: Aug. 10, 2001
Returned: Aug. 22, 2001

STS-108 Endeavour

Daniel Tani, U.S.
Linda Godwin, U.S.
Dominic Gorie, U.S.
Mark Kelly, U.S.
Daniel Bursch, U.S., up*
Yuri Onufrienko, Russia (Roscosmos), up*
Carl Walz, U.S., up*
Frank Culbertson, U.S., down*
Vladimir Dezhurov, Russia (Roscosmos), down*
Mikhail Turin, Russia (Roscosmos), down*
Launched: Dec. 5, 2001
Returned: Dec. 17, 2001

STS-110 Atlantis

Michael Bloomfield, U.S.
Stephen Frick, U.S.
Lee Morin, U.S.
Ellen Ochoa, U.S.
Jerry Ross, U.S.
Steven Smith, U.S.
Rex Walheim, U.S.
Launched: Apr. 8, 2002
Returned: Apr. 19, 2002

STS-111 Endeavour

Franklin Chang-Diaz, U.S.
Kenneth Cockrell, U.S.
Paul Lockhart, U.S.
Philippe Perrin, France (CNES)
Valery Korzun, Russia (Roscosmos), up*
Sergei Treschev, Russia (Roscosmos), up*
Peggy Whitson, U.S., up*
Daniel Bursch, U.S., down*
Yuri Onufrienko, Russia (Roscosmos), down*
Carl Walz, U.S., down*
Launched: June 5, 2002
Returned: June 19, 2002

* "Up" means that the crewmember launched on this flight; "down" means that the crewmember returned on this flight.

(continued on page 86)

STS Missions and Crews

(continued from page 85)

Mission	Patch	Crew
STS-112		
STS-113		
STS-114		
STS-121		
STS-115		

STS-112 Atlantis
Jeffrey Ashby, U.S.
Sandra Magnus, U.S.
Pamela Melroy, U.S.
Piers Sellers, U.S.
David Wolf, U.S.
Fyodor Yurchikhin, Russia (Roscosmos)
Launched: Oct. 7, 2002
Returned: Oct. 18, 2002

STS-113 Endeavour
John Herrington, U.S.
Paul Lockhart, U.S.
Michael Lopez-Alegria, U.S.
James Wetherbee, U.S.
Kenneth Bowersox, U.S., up*
Nikolai Budarin, Russia (Roscosmos), up*
Donald Pettit, U.S., up*
Valery Korzun, Russia (Roscosmos), down*
Sergei Treschev, Russia (Roscosmos), down*
Peggy Whitson, U.S., down*
Launched: Nov. 23, 2002
Returned: Dec. 7, 2002

STS-114 Discovery
Eileen Collins, U.S.
James Kelly, U.S.
Soichi Noguchi, Japan (JAXA)
Stephen Robinson, U.S.
Andrew Thomas, U.S.
Wendy Lawrence, U.S.
Charles Camarda, U.S.
Launched: July 26, 2005
Returned: Aug. 9, 2005

STS-121 Discovery
Steven Lindsey, U.S.
Mark Kelly, U.S.
Michael Fossum, U.S.
Piers Sellers, U.S.
Lisa Nowak, U.S.
Stephenie Wilson, U.S.
Thomas Reiter, Germany (ESA), up*
Launched: July 4, 2006
Returned: July 17, 2006

STS-115 Atlantis
Brent Jett, U.S.
Christopher Ferguson, U.S.
Heidemarie Stefanyshyn-Piper, U.S.
Joseph Tanner, U.S.
Daniel Burbank, U.S.
Steven MacLean, Canada (CSA)
Scheduled to launch: Aug. 2006
Scheduled to return: Sept. 2006

* "Up" means that the crewmember launched on this flight; "down" means that the crewmember returned on this flight.

Soyuz docked to the FGB. The Space Shuttle is in the background.

Shuttle ISS Missions

Shuttle Flight/ ISS Sequence No.	Launched	Landed	Docked
STS-88/2A	12/04/98	12/15/98	6 d, 20 h, 38 m
STS-96/2A.1	05/27/99	06/06/99	5 d, 18 h, 17 m
STS-101/2A.2a	05/19/00	05/29/00	5 d, 18 h, 32 m
STS-106/2A.2b	09/08/00	09/19/00	7 d, 21 h, 54 m
STS-92/3A	10/11/00	10/24/00	6 d, 21 h, 24 m
STS-97/4A	11/30/00	12/11/00	6 d, 23 h, 13 m
STS-98/5A	02/07/01	02/20/01	6 d, 21 h, 15 m
STS-102/5A.1	03/08/01	03/21/01	8 d, 21 h, 54 m
STS-100/6A	04/19/01	05/01/01	8 d, 3 h, 35 m
STS-104/7A	07/12/01	07/24/01	8 d, 1 h, 46 m
STS-105/7A.1	08/10/01	08/22/01	7 d, 20 h, 10 m
STS-108/UF-1*	12/05/01	12/17/01	7 d, 21 h, 25 m
STS-110/8A	04/08/02	04/19/02	7 d, 2 h, 26 m
STS-111/UF-2*	06/05/02	06/19/02	7 d, 22 h, 7 m
STS-112/9A	10/07/02	10/18/02	6 d, 21 h, 56 m
STS-113/11A	11/23/02	12/07/02	6 d, 22 h, 6 m
STS-114/LF-1*	07/26/05	08/09/05	8 d, 19 h, 54 m
STS-121/ULF-1.1*	07/01/06	07/17/06	8 d, 19 h, 16 m
TOTALS:			134 d, 9 h, 48 m

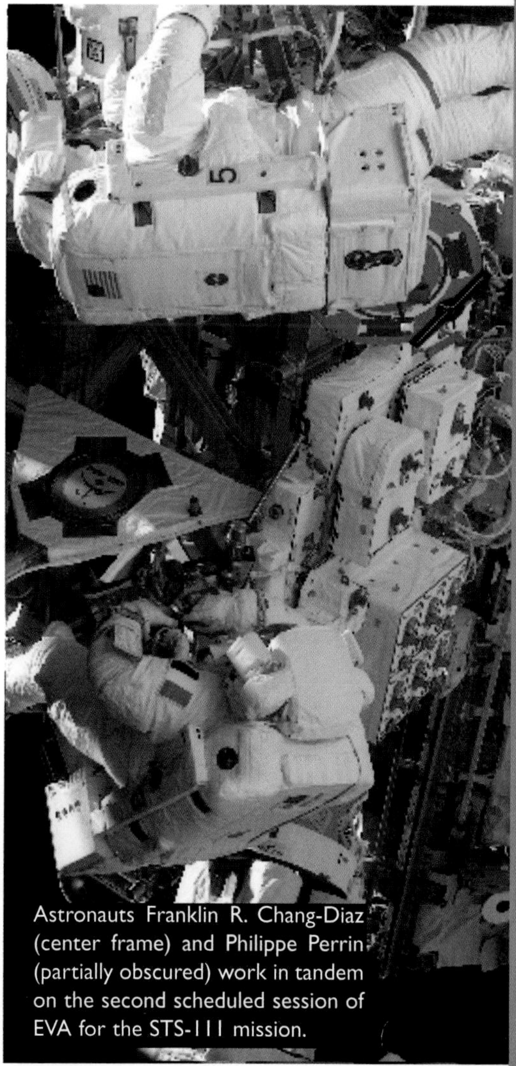
Astronauts Franklin R. Chang-Diaz (center frame) and Philippe Perrin (partially obscured) work in tandem on the second scheduled session of EVA for the STS-111 mission.

Astronaut Michael L. Gernhardt participates in a spacewalk aimed toward wrapping up work on the second phase of the ISS.

Soyuz ISS Missions

Spacecraft	Launched	Duration (Docked)	Landed	Crew Up	Crew Down
Soyuz TM-31 1S	10/31/00	186 days	05/06/01	Yuri Gidzenko, Russia (Roscosmos), Sergei Krikalev, Russia (Roscosmos), William Shepherd, USA (NASA)	Talgat Musabayev, Russia (Roscosmos), Yuri Baturin, Russia (Roscosmos), Dennis Tito, USA, SFP†
Soyuz TM-32 2S	04/28/01	186 days	10/31/01	Talgat Musabayev, Russia (Roscosmos), Yuri Baturin, Russia (Roscosmos), Dennis Tito, USA, SFP†	Viktor Afansayev, Russia (Roscosmos), Claudie Hagniere, France (CNES), Konstantin Kozayev, Russia (Roscosmos)
Soyuz TM-33 3S	10/21/01	196 days	05/05/02	Viktor Afansayev, Russia (Roscosmos), Claudie Hagniere, France (CNES), Konstantin Kozayev, Russia (Roscosmos)	Yuri Gidzenko, Russia (Roscosmos), Roberto Vittori, Italy (ESA), Konstantin Kozayev, Russia (Roscosmos),
Soyuz TM-34 4S	04/25/02	198 days	11/10/02	Yuri Gidzenko, Russia (Roscosmos), Roberto Vittori, Italy (ESA), Konstantin Kozayev, Russia (Roscosmos)	Sergei Zalyotin, Russia (Roscosmos), Frank De Winne, Belgium (ESA), Yuri Lonchakov, Russia (Roscosmos)
Soyuz TMA-1 5S	10/30/02	186 days	05/04/03	Sergei Zalyotin, Russia (Roscosmos), Frank De Winne, Belgium (ESA), Yuri Lonchakov, Russia (Roscosmos)	Nikolai Budarin, Russia (Roscosmos) Kenneth Bowersox, U.S. (NASA) Donald Pettit, U.S. (NASA)
Soyuz TMA-2 6S	04/26/03	185 days	10/28/03	Yuri Malenchenko, Russia (Roscosmos) Edward Lu, U.S. (NASA)	Yuri Malenchenko, Russia (Roscosmos) Edward Lu, U.S. (NASA) Pedro Duque, Spain (ESA)
Soyuz TMA-3 7S	10/18/03	192 days	04/30/04	Michael Foale, U.S. (NASA) Alexander Kaleri, Russia (Roscosmos) Pedro Duque, Spain (ESA)	Michael Foale, U.S. (NASA) Alexander Kaleri, Russia (Roscosmos) Andre Kuipers, Netherlands (ESA)
Soyuz TMA-4 8S	04/19/04	187 days	10/24/04	Gennady Padalka, Russia (Roscosmos) Edward Michael Fincke, U.S. (NASA) Andre Kuipers, Netherlands (ESA)	Gennady Padalka, Russia (Roscosmos) Edward Michael Fincke, U.S. (NASA) Yuri Shargin, Russia (Roscosmos)
Soyuz TMA-5 9S	10/14/04	193 days	04/24/05	Salizhan Sharipov, Russia (Roscosmos) Leroy Chiao, U.S. (NASA) Yuri Shargin, Russia (Roscosmos)	Salizhan Sharipov, Russia (Roscosmos) Leroy Chiao, U.S. (NASA) Roberto Vittori, Italy (ESA)
Soyuz TMA-6 10S	04/15/05	180 days	10/11/05	Sergei Krikalev, Russia (Roscosmos) John Phillips, U.S. (NASA) Roberto Vittori, Italy (ESA)	Sergei Krikalev, Russia (Roscosmos) John Phillips, U.S. (NASA) Gregory Olsen, U.S., SFP†
Soyuz TMA-7 11S	10/01/05	190 days	04/08/06	Valery Tokarev, Russia (Roscosmos) William McArthur, U.S. (NASA) Gregory Olsen, U.S., SFP†	Valery Tokarev, Russia (Roscosmos) William McArthur, U.S. (NASA) Marcos Pontes, Brazil, SFP†
Soyuz TMA-8 12S	03/30/06	178 days planned	09/25/06 planned	Pavel Vinogradov, Russia (Roscosmos) Jeffrey Williams, U.S. (NASA) Marcos Pontes, Brazil, SFP†	Pavel Vinogradov, Russia (Roscosmos) Jeffrey Williams, U.S. (NASA) TBD

† Space Flight Participant (SFP)

Progress ISS Missions

Spacecraft*	ISS Flight Sequence	Launched	Undocked	Duration (Docked)	Deorbit
Progress M1-3	1P	08/06/00	11/01/00	84 d, 7 h, 51 m	11/01/00
Progress M1-4	2P	11/16/00	02/08/01	82 d, 7 h, 39 m	02/08/01
Progress M-44	3P	02/26/01	04/13/01	46 d, 22 h, 58 m	04/13/01
Progress M1-6	4P	05/21/01	08/22/01	91 d, 5 h, 38 m	08/22/01
Progress M-45	5P	08/21/01	11/22/01	91 d, 6 h, 21 m	11/22/01
Progress M1-7	6P	11/26/01	03/19/02	110 d, 22 h	03/20/02
Progress M1-8	7P	03/21/02	06/25/02	92 d, 11 h, 29 m	06/25/02
Progress M-46	8P	06/26/02	09/24/02	87 d, 7 h, 36 m	10/14/02
Progress M1-9	9P	09/25/02	02/01/03	124 d, 23 h	02/01/03
Progress M-47	10P	02/02/03	08/28/03	205 d, 7 h, 59 m	08/28/03
Progress M1-10	11P	06/08/01	09/04/03	85 d, 8 h, 26 m	10/03/03
Progress M-48	12P	08/29/03	01/28/04	150 d, 4 h, 55 m	01/29/04
Progress M1-11	13P	01/29/04	05/24/04	113 d, 20 h, 06 m	06/03/04
Progress M-49	14P	05/25/04	07/30/04	63 d, 16 h, 11 m	07/30/04
Progress M-50	15P	08/11/04	12/22/04	130 d, 13 h, 34 m	12/22/04
Progress M-51	16P	12/23/04	02/27/05	65 d	03/09/05
Progress M-52	17P	02/28/05	06/16/05	106 d, 1 h, 5 m	06/16/05
Progress M-53	18P	06/17/05	09/07/05	80 d, 5 h, 41 m	09/07/05
Progress M-54	19P	09/08/05	03/03/06	173 d, 23 h, 24 m	03/05/06
Progress M-55	20P	12/21/05	6/19/06	179d	06/20/06
Progress M-56	21P	04/24/06	Planned 09/15/06	TBD	TBD
Progress M-57	22P	06/24/06	Planned 12/19/06	TBD	TBD

* All Progress spacecraft were/will be launched by Soyuz launch vehicles.

Interesting Points

- The ISS effort involves more than 100,000 people in space agencies, at 500 contractor facilities, and in 37 U.S. states. That's almost half of the entire population of the U.S. state of North Dakota.
- As of June 2006, the number of crewmembers and visitors who have traveled to the ISS included 116 different people representing 10 countries.
- Living and working on the ISS is like building one room of a house, moving in a family of three, and asking them to finish building the house while working full time from home.
- As of June 2006:
  - Including the launch of the first module—Zarya at 1:40 a.m. e.s.t. on November 20, 1998—there have been 55 launches to the ISS (37 Russian flights and 18 U.S./Shuttle flights).
  - The 38 Russian flights include 3 modules (Zarya, Zvezda, and Pirs), 13 Soyuz crew vehicles, and 22 Progress resupply ships.
- At Assembly Complete, 80 space flights will have been scheduled to take place using five different types of launch vehicles.

EVA

- As of August 2006:
  - Spacewalks (EVAs): 69 (28 Shuttlebased, 41 ISS-based) totaling 410 hours.
- Building the ISS in space has been compared to changing a spark plug or hanging a shelf while wearing roller skates and two pairs of ski gloves with all your tools, screws, and materials tethered to your body so they don't drop.

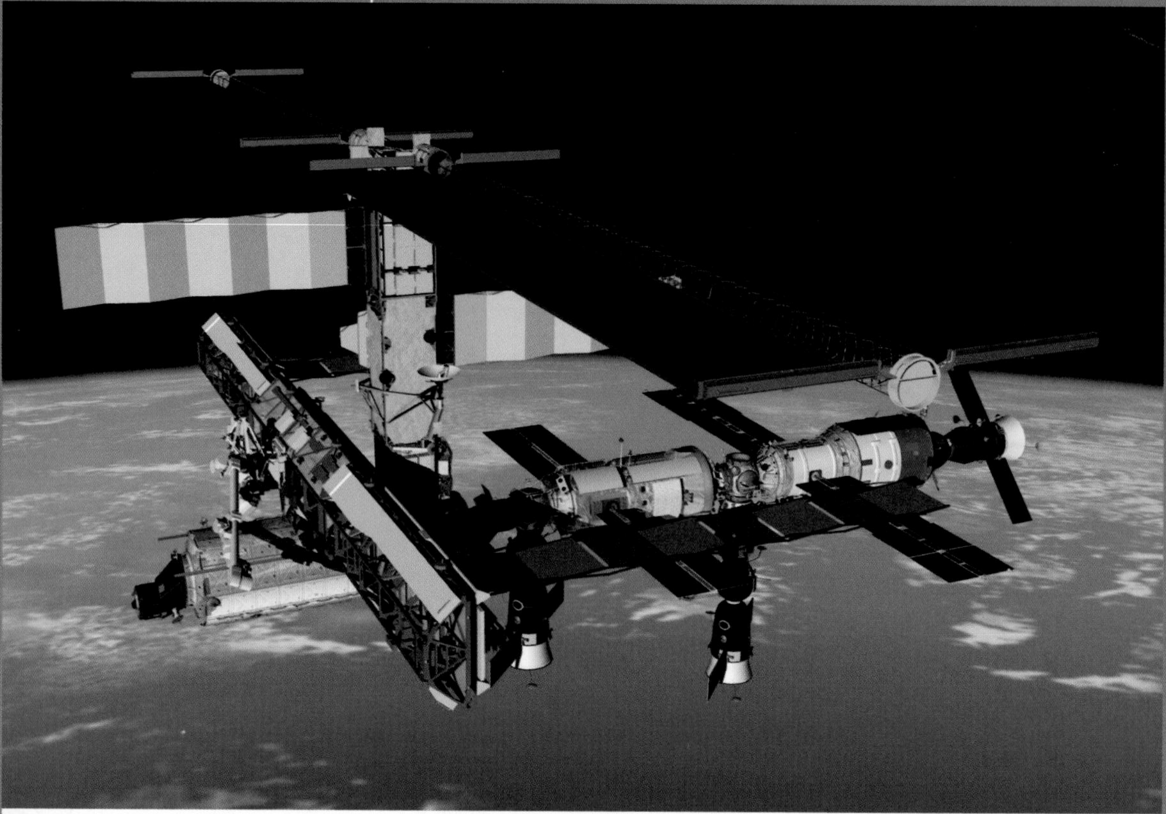

Physical Parameters

Mass

- The mass of the ISS currently is 186,000 kg (410,000 lb) (equivalent to about 132 automobiles).
- At Assembly Complete, the ISS will be about four times as large as the Russian space station Mir and about five times as large as the U.S. Skylab.
- At Assembly Complete, the ISS will have a mass of almost 419,600 kg (925,000 lb). That's the equivalent of more than 330 automobiles.
- The entire 16.4-m (55-ft) robot arm assembly will be able to lift 99,790 kg (220,000 lb), which is the mass of a Space Shuttle orbiter.

Habitable Volume

- The ISS has about 425 m^3 (15,000 ft^3) of habitable volume—more room than a conventional three-bedroom house. There are 9 research racks on board plus 16 system racks and 10 stowage racks.
- At Assembly Complete, more than 120 telephone-booth-size rack facilities will be installed in the ISS for operating the spacecraft systems and research experiments.
- When completely assembled, the ISS will have an internal pressurized volume of 935 m^3 (33,023 ft^3), or about 1.5 Boeing 747s, and will be larger than a five-bedroom house.

Physical Dimensions

- The ISS solar array surface will be large enough to cover the U.S. Senate Chamber more than three times over at Assembly Complete.
- A solar array's wingspan of 73 m (240 ft) is longer than that of a Boeing 777, which is 65 m (212 ft).
- At Assembly Complete, the ISS will measure 110 m (361 ft) end to end. That's equivalent to the length of a U.S. football field, including the end zones.

Electrical Power

- The solar array surface area currently on orbit is 892 m^2 (9,600 ft^2), which is large enough to cover 75% of the U.S. House of Representatives Chamber (42 m x 28 m = 1,176 m^2) (139 ft x 93 ft = 12,927 ft^2).
- At Assembly Complete, 12.9 km (8 mi) of wire will connect the electrical power system.
- Currently, 26 kW of power is generated.
- At Assembly Complete, the solar array surface area is 2,500 m^2 (27,000 ft^2), an acre of solar panels.
- At Assembly Complete, there will be a total of 262,400 solar cells.
- At Assembly Complete, a maximum 110 kW of power, including 30 kW of long-term average power for applications, is/will be available.

Thermal Control

- Currently, there are 21 honeycombed aluminum radiator panels, each measuring 1.8 m x 3 m (6 ft x 10 ft), for a total of 156 m^2 (1,680 ft^2) of ammoniatubing-filled heat exchange area.
- At Assembly Complete, there will be 42 honeycombed aluminum radiator panels, each measuring 1.8 m x 3 m (6 ft x 10 ft), for a total area of 312 m^2 (3,360 ft^2) of ammonia-tubing-filled heat exchange area.

Module Berthing

- To ensure a good seal, the Common Berthing Mechanism automatic latches pull two modules together and tighten 16 connecting bolts with a force of 8,618 kg (19,000 lb) each.

Meals

- Crews have eaten about 23,000 meals and 20,000 snacks, which equals 18,150 kg (40,000 lb) of food. Approximately 3,630 kg (4 tons) of supplies are required to support a crew of three for about 6 months.
- Based on input from ISS crew members, the most popular on-orbit foods are shrimp cocktail, tortillas, barbecue beef brisket, breakfast sausage links, chicken fajitas, vegetable quiche, macaroni and cheese, candy-coated chocolates, and cherry blueberry cobbler. The favorite beverage to wash it all down? Lemonade.

Crew Hours

- While a year of Space Shuttle operations (seven crew members, 11-day missions, five flights per year) results in 9,240 total crew hours, 1 year of ISS operations—26,280 total crew hours (three crew, 365 days)—is almost three times that amount.

Environmental Control

- ISS systems recycle about 6.4 kg (14 lb) or 6.42 L (1.7 gal) of crew-expelled air each day. 2.7 kg (6 lb) of that comes from the U.S. segment. The processed water is then used for technical or drinking purposes.
- The ISS travels an equivalent distance to the Moon and back in about a day. That's equivalent to crossing the North American continent about 135 times every day.

Data Management

- Fifty-two computers will control the systems on the ISS.
- The data transmission rate is 150 Mb per second downlink with simultaneous uplink.
- Currently, 2.8 million lines of software code on the ground will support 1.5 million lines of flight software code, which will double by Assembly Complete.

- In the International Space Station's U.S. segment alone, 1.5 million lines of flight software code will run on 44 computers communicating via 100 data networks transferring 400,000 signals (e.g., pressure or temperature measurements, valve positions, etc.).
- The ISS will manage 20 times as many signals as the Space Shuttle.

Research and Applications

- Expedition crews conduct science daily, across a wide variety of fields, including human research, life sciences, physical sciences, and Earth observation, as well as education and technology demonstrations (http://exploration.nasa.gov/programs/station).
- As of June 2006, 90 science investigations have been conducted on the ISS over 64 months of continuous research. Nine research racks are on board. More than 7,700 kg (17,000 lb) of research equipment and facilities have been brought to the ISS.
- Research topics have been diverse—from protein crystal growth to physics to telemedicine. New scientific results from early Space Station research, in fields from basic science to exploration research, are being published every month.
- Some 100 scientists, from as many institutions, have been principal investigators on ISS research, either completed or ongoing. NASA research has involved lead investigators from the U.S., Belgium, Canada, France, Germany, Italy, Japan, the Netherlands, and Spain. On some experiments, these principal investigators represent dozens of scientists who share data to maximize research.
- The ISS provides an excellent viewing platform for Earth; its range covers more than 90% of the populated areas of the planet. Station crews have taken more than 200,000 images of Earth—almost a third of the total number of images taken from orbit by astronauts.
- About 700,000 NASA digital photographs of Earth are downloaded by scientists, educators, and the public each month from the "Gateway to Astronaut Photography of Earth" (http://eol.jsc.nasa.gov).
- In 2005, ISS astronauts took key photographs of the hurricane damage in Mississippi and Louisiana, as well as damage and recovery efforts from the tsunami in Sri Lanka; documented floods and droughts; and took detailed photographs of cities around the world, from London to Jeddah to Irkutsk.

Education

- Educational activities relating to the ISS include student-developed experiments; educational demonstrations and activities; and student participation in classroom versions of ISS experiments, NASA investigator experiments, and ISS engineering activities.
- From early 2000 through April 2006, 24 unique types of educational programs involved 31.8 million students, and over 12,500 teachers participated in ISS-based education workshops.
- In the EarthKAM experiment, nearly 1,000 schools and 66,000 middle school students have controlled a digital camera on board the ISS to photograph features of Earth. The students have investigated a wide range of topics such as deforestation, urbanization, volcanoes, river deltas, and pollution.
- In-flight education downlinks (part of Education Payload Operations) have linked crewmembers aboard the ISS with students around the world. The students have studied the science activities on the ISS and living and working in space in preparation for asking questions of the crewmembers. Through broadcasts sponsored by Channel One and the U.S. Department of Education, over 30 million students have been able to watch the interviews.

Crew Medical Care

- Information from biomedical research on ISS is designed to develop countermeasures to the negative effects of longduration space flight on the human body so that future astronauts will be able to explore more safely. For example,
- Resistive exercise allows astronauts to do weight training while they are weightless and is being studied to see if it can slow the rate of bone loss that occurs in space.
- Genetic techniques will soon be used to examine the microbial environment of the Space Station, and culture studies will determine the effect of the space environment on the growth of microbes. This will allow better assessment of the risks of pathogens to crewmembers on long-duration missions.
- Medical ultrasound will be used as a diagnostic tool should a crewmember be hurt, even if the rest of the crew has not been previously trained in how to do a specific type of scan. The same telemedicine techniques benefit patients in rural areas and may eventually allow ultrasound images taken on ambulances to be sent ahead to the hospital.

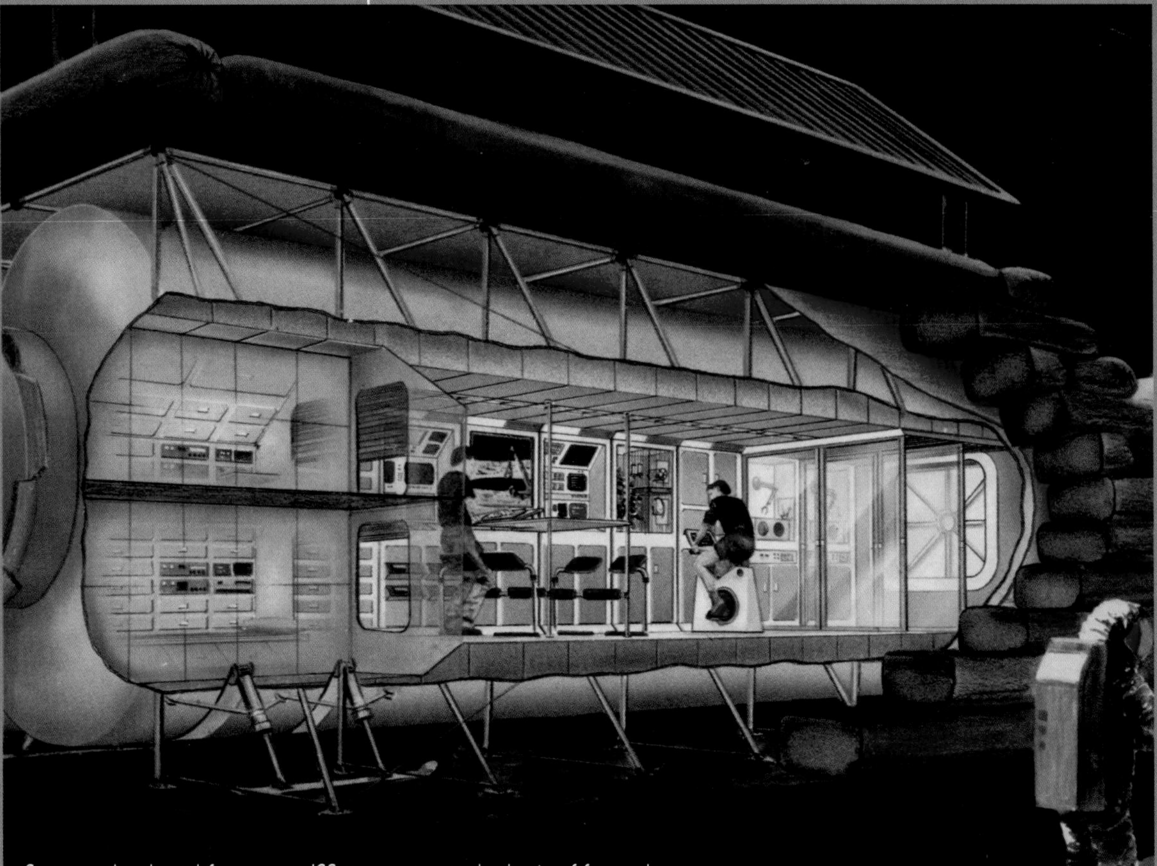

Systems developed for use on ISS may serve as the basis of future lunar outposts.

The International Space Station (ISS) is instrumental to the exploration of space.

Efficient, reliable spacecraft systems are critical to reducing crew and mission risks. The development and testing of systems of the ISS will reduce mission risks and advance capabilities for missions traveling interplanetary distances.

As we expand permanent human presence beyond low-Earth orbit to the Moon and, later, to Mars and beyond, we will face challenges in management; integration; remote, long-duration assembly and maintenance operations; science and engineering; and international culture and relationships. The ISS Program is providing critical insight and amassing new knowledge in all of these areas, and the ISS experience can help to guide our success in space exploration.

appendix

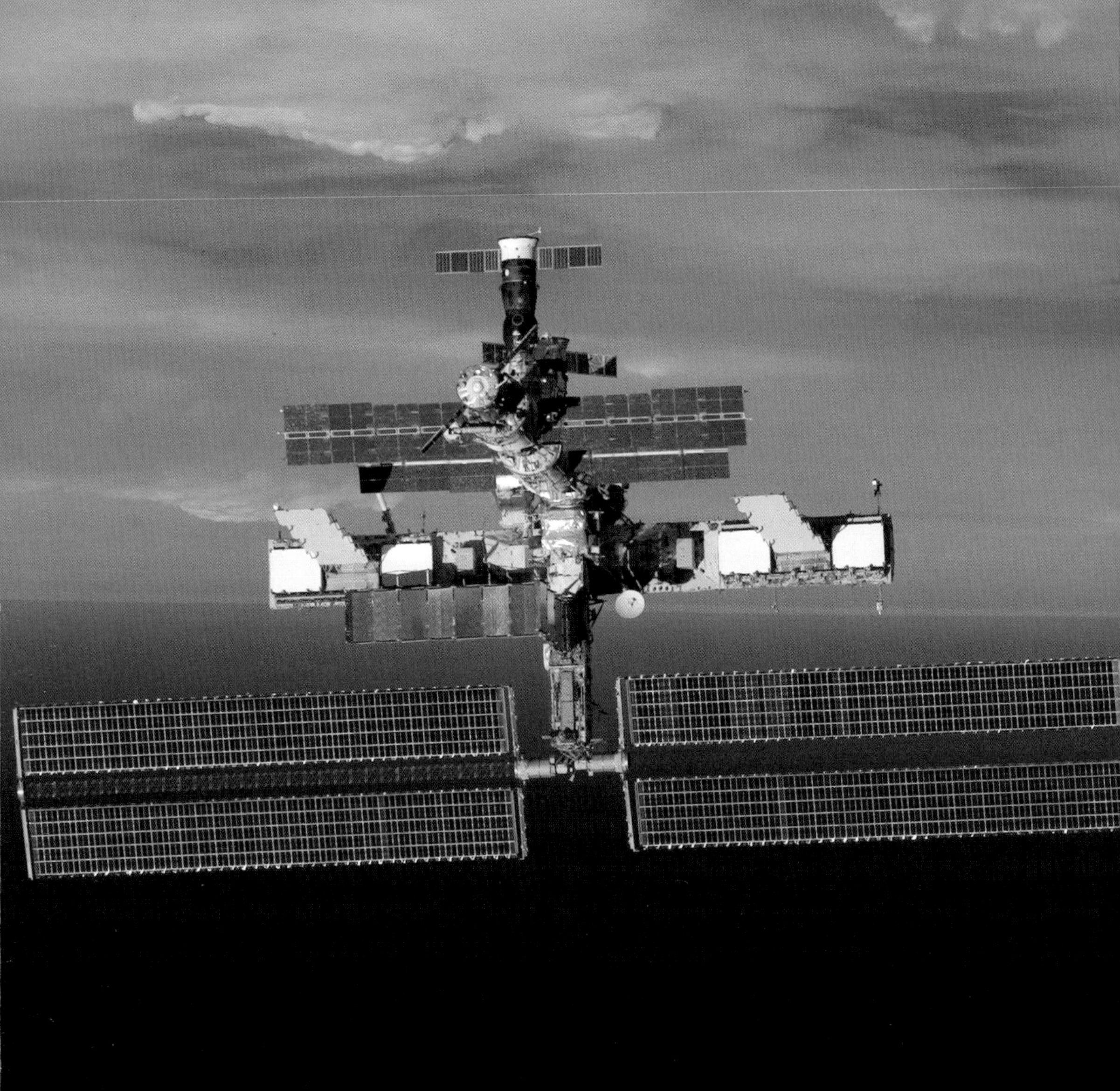

NASA wishes to acknowledge the use of images provided by:

Canadian Space Agency

European Space Agency

Japan Aerospace Exploration Agency

Roscosmos, the Russian Federal Space Agency

Acronym List

1P	Progress flight
1S	Soyuz flight
AC	Assembly Complete
ACU	Arm Control Unit
ARC	Ames Research Center
ARIS	Active Rack Isolation System
ATCS	Active Thermal Control System
atm	Atmospheres
ATV	Automated Transfer Vehicle, launched by Ariane [ESA]
ATV-CC	Automated Transfer Vehicle Control Centre
BCA	Battery Charging Assembly
BCDU	Battery Charge Discharge Unit
BSA	Battery Stowage Assembly
CBM	Common Berthing Mechanism
CC	Control Center
CCAA	Common Cabin Air Assembly
CCC	Contaminant Control Cartridge
CDRA	Carbon Dioxide Removal Assembly
CETA	Crew and Equipment Translation Aid/Assembly
CEV	Crew Exploration Vehicle
CEVIS	Cycle Ergometer with Vibration Isolation System
CHeCS	Crew Health Care System
CMG	Control Moment Gyroscope
CMRS	Crew Medical Restraint System
CMS	Countermeasures System
CNES	Centre National D'Études Spatiales [French space agency]
COF	Columbus Orbital Facility
COL-CC	Columbus Control Centre
COTS	Commercial Orbital Transportation Services
CPDS	Charged Particle Directional Spectrometer
CRPCM	Canadian Remote Power Controller Module
CSA	Canadian Space Agency
CTB	Cargo Transfer Bag
CWC	Contingency Water Container
DC	Docking Compartment; Direct Current
DCSU	Direct Current Switching Unit
DDCU	DC-to-DC Converter Unit
DDT&E	Design, Development, Test, and Evaluation
DLR	German Aerospace Center
DMS	Data Management System
DOS	Long-Duration Orbital Station [Russian]
EADS	European Aeronautic Defence and Space Company
ECLSS	Environmental Control and Life Support System
ECS	Exercise Countermeasures System
ECU	Electronics Control Unit
EDR	European Drawer Rack
EDV	Water Storage Container [Russian]
EF	Exposed Facility
EHS	Environmental Health System
ELC	ExPRESS Logistics Carrier
ELM	Experiment Logistics Module
EMU	Extravehicular Mobility Unit
EPM	European Physiology Module
EPS	Electrical Power System
ERA	European Robotic Arm
ESA	European Space Agency
ESTEC	European Space Research and Technology Centre
ETC	European Transport Carrier
EVA	Extravehicular Activity
ExPCA	ExPRESS Carrier Avionics
EXPRESS	Expedite the Processing of Experiments to the Space Station
FGB	Functional Cargo Block
FRAM	Flight Releasable Attachment Mechanism
FRGF	Flight Releasable Grapple Fixture
FSA	Roscosmos, Russian Federal Space Agency
FSL	Fluid Science Laboratory
GASMAP	Gas Analyzer System for Metabolic Analysis Physiology
GB	Gigabyte
GCM	Gas Calibration Module
GCTC	Gagarin Cosmonaut Training Center
GN&C	Guidance, Navigation, and Control
GLONASS	Global Navigation Satellite System [Russian]
GPS	Global Positioning System
GRC	Glenn Research Center
GSC	Guiana Space Center
HMS	Health Maintenance System
HRF	Human Research Facility
HTV	H-II Transfer Vehicle [JAxA]
IBMP	Institute for Biomedical Problems
ICC	Integrated Cargo Carrier
ICS	Internal Communications System
IEA	Integrated Equipment Assembly
IRU	In-flight Refill Unit
ISPR	International Standard Payload Rack
ISS	International Space Station
ITA	Integrated Truss Assembly
ITS	Integrated Truss Structure
IV-CPDS	Intravehicular Charged Particle Directional Spectrometer
JAXA	Japan Aerospace Exploration Agency
JEM	Japanese Experiment Module
JEM-ELM	Japanese Experiment Module-Experiment Logistics Module
JEM-ELM-EF	Japanese Experiment Module-Experiment Logistics Module-Exposed Facility
JEM-ELM-ES	Japanese Experiment Module-Experiment Logistics Module-Exposed Section
JEM-ELM-PS	Japanese Experiment Module-Experiment Logistics Module-Pressurized Section
JEM-PM	Japanese Experiment Module-Pressurized Module
JEM-RMS	Japanese Experiment Module-Remote Manipulator System
JSC	Johnson Space Center
kgf	Kilogram Force
kN	Kilonewton
KSC	Kennedy Space Center
lbf	Pound Force
LF	Logistics Flight
LiOH	Lithium Hydroxide
LSS	Life Support Subsystem
Mb	Megabit
MBS	Mobile Base System
MBSU	Main Bus Switching Unit
MCC	Mission Control Center
MDM	Multiplexer-Demultiplexer
MELFI	Minus Eighty-Degree Laboratory Freezer for ISS
MGBX	Microgravity Science Glovebox
MLE	Middeck Locker Equivalent
MLM	Multipurpose Laboratory Module
MMOD	Micrometeoroid/Orbital Debris

(continued on page 102)

Acronym List

(continued from page 101)

MMU	Mass Memory Unit
MOC	MSS Operations Complex
MPLM	Multi-Purpose Logistics Module
MSFC	Marshall Space Flight Center
MSS	Mobile Servicing System
MT	Mobile Transporter
NASA	National Aeronautics and Space Administration
NAVSTAR	Navigation Signal Timing and Ranging [U.S. satellite]
NPO	Production Enterprise [Russian]
NTO	Nitrogen Tetroxide
NTSC	National Television Standards Committee
OMS	Orbital Maneuvering System
OGS	Oxygen Generation System
ORU	Orbital Replacement Unit
OVC	Oxygen Ventilation Circuit
P1, P6, etc.	Port trusses
PCAS	Passive Common Attach System
PDA	Payload Disconnect Assembly
PDGF	Payload Data Grapple Fixture
PLSS	Primary Life Support System
PM	Pressurized Module
PMA	Pressurized Mating Adapter
POC	Payload Operations Center; Primary Oxygen Circuit
PROX OPS	Proximity Operations
PSA	Power Supply Assembly
PSC	Physiological Signal Conditioner
PTCS	Passive Thermal Control System
PVGF	Power Video Grapple Fixture
PVR	Photovoltaic Radiator
RED	Resistive Exercise Device
RGA	Rate Gyro Assembly
RM	Research Module
RMS	Remote Manipulation, Manipulator System
RPC	Remote Power Controller
rpm	Revolutions Per Minute
ROEU-PDA	Remotely Operated Electrical Umbilical-Power Distribution Assembly
RPCM	Remote Power Controller Module
RSC	Rocket and Space Corporation
RV	Reentry Vehicle
S&M	Structures and Mechanisms
S0 or S	Zero, Starboard trusses
S1, etc.	
SARJ	Solar (Array) Alpha Rotation Joint
SAFER	Simplified Aid for EVA Rescue
SASA	S-Band Antenna Structural Assembly
SAW	Solar Array Wing
SFOG	Solid Fuel Oxygen Generator
SFP	Space Flight Participant
SGANT	Space-to-Ground Antenna
SM	Service Module
SPDM	Special Purpose Dexterous Manipulator
SS	Space Shuttle
SSA	Space Suit Assembly
SSIPC	Space Station Integration and Promotion Center
SSRMS	Space Station Remote Manipulator System
SSU	Sequential Shunt Unit
STS	Space Transportation System
TCS	Thermal Control System
TDRS	Tracking and Data Relay Satellite
TEPC	Tissue Equivalent Proportional Counter
TKS	Orbital Transfer System
TKSC	Tsukuba Space Center
TMA	Transportation Modified Anthropometric
TMG	Thermal Micrometeoroid Garment
TNSC	Tanegashima Space Center
TORU	Progress Remote Control Unit [Russian]
TSC	Telescience Support Center
TSS	Temporary Sleep Station
TSUP	Moscow Mission Control
TVIS	Treadmill Vibration Isolation System
UDMH	Unsymmetrical Dimethylhydrazine
UF	Utilization Flight
UHF	Ultra-High Frequency
ULF	Utilization and Logistics Flight
UMA	Umbilical Mating Assembly
USOC	User Support and Operations Centre
VDC	Voltage, Direct Current
VDU	Video Distribution Unit
VOA	Volatile Organic Analyzer
WRS	Water Recovery System
Z1	Zenith 1, a truss segment

Definitions

Assembly Complete:
Final integrated arrangement of ISS elements

Assembly Stage:
Integrated arrangement of ISS elements

Berthing:
Linking of two spacecraft, modules, or elements; uses apparatus with wide internal hatch

Docking:
Linking of two spacecraft; uses apparatus with narrow internal hatch

Element:
A structural component such as a module or truss segment

Expedition:
A stay on board the ISS; the long-duration crew during a stay on the ISS

Increment:
Period of time from launch of a vehicle rotating ISS crewmembers to the undocking of the return vehicle for that crew

Mission:
Flight of a "visiting" Space Shuttle, Soyuz, or other vehicle not permanently attached to the ISS

Module:
An internally pressurized element intended for habitation

Multiplexer:
A computer that interleaves multiple data management functions

Nadir:
Directly below, opposite zenith

Port:
Left side, opposite starboard

Rendezvous:
Movement of two spacecraft toward one another

Space Flight Participant:
Nonprofessional astronaut

Starboard:
Right side, opposite port

Zenith:
Directly above, opposite nadir

ISS Partners:

United States of America
www.nasa.gov

Canada
www.space.gc.ca/asc/eng/default.asp

Japan
www.jaxa.jp/index_e.html

Russian Federation
www.roscosmos.ru

European Space Agency
www.esa.int

APOLLO 11
FIRST MEN ON THE MOON
Robert Godwin
POCKET SPACE GUIDE

MARS
Robert Godwin
POCKET SPACE GUIDE

DEEP SPACE
Stephen Whitfield
POCKET SPACE GUIDE

LAUNCH
Vehicles
Heritage of the Space Race
Michael Lennick
POCKET SPACE GUIDE

APOLLO
The Test Program
PROJECT
Robert Godwin
POCKET SPACE GUIDE

APOLLO
Exploring the Moon
PROJECT
Robert Godwin
POCKET SPACE GUIDE

HUBBLE
Space Telescope
Steve Whitfield
POCKET SPACE GUIDE

RUSSIAN
Spacecraft
Robert Godwin
POCKET SPACE GUIDE

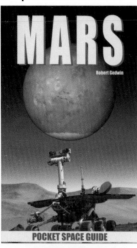

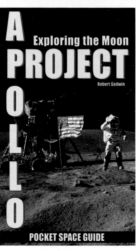